Chemical Signals

A novel by:

J.C. Rivers

Chemical Signals by J.C. Rivers

ISBN: 979-8-9867117-2-0

United States of America

Flow State Publishers, LLC

Cover art by Cia Lindgren.

FOR MY FAMILY

CHEMICAL SIGNATURES

It is a little-known fact that each tree is unique and has a distinct chemical signal. This unique signal allows forensics experts to trace them to their point of geographic origin. Illegal timber has payouts in the millions to billions of dollars and the fallout can mean ecological catastrophe.

CLOSE CALL

"Shhh, I think we just need to go a little farther," Clara whispered to Marcel, holding his hand as they crept down a game trail that led out from the pull-off where they parked up the road.

"Why are we being so secretive, Clara? This is a business. Can't you just go there and ask them your questions like a normal person?" Marcel asked, his tone half laughing, half annoyed.

"If there are nefarious things happening here, then we won't see them if we announce we are coming, now will we?" She shot him a mischievous smile.

Clara was whimsical and loved adventure. She was a budding investigative journalist with just enough street savvy to keep her safe, and just a little too much trust in the world. Marcel was also studying journalism, but his beat was going to be humanitarian crises and war zones; he also loved a good adventure.

"It's like we are practicing for the real thing. One day, when we have a big story, we may have to actually do stuff like this," said Clara as the stacks of logs came into sight.

"There it is," she whispered. "Wow, this looks pretty high-tech for tree stuff. I guess I didn't expect to see so much equipment."

"And so much security. Guess they do have some pretty expensive trees in there."

Clara heard that a new mill operation opened a little less than a year ago. They were making high-end finished products from hardwood trees, mostly from the Appalachian Mountains and other Kentucky groves. Rumor had it that they were making products for celebrities and high-end buyers. Clara was curious what would draw in that kind of clientele. Plus, it was a great way to spend a Tuesday afternoon. Way better than being at home studying.

Just then, an Audi pulled up the dirt road, rang the intercom, and waited for the lot gate to swing open. It looked like there were three men in the car, but the windows were tinted so it was hard to tell. A man came out from inside the main building to greet them. One man stepped out from the passenger side. Then the driver got out and stood by the back door of the car. Clara strained to hear what they were saying, but they were too far away. It looked heated. In a flash, guns were pulled.

"Holy crap, this just got really interesting. I think we should get out of here, Clara. Whatever is going on here, we don't want any part of it," said Marcel.

"Shh, just another minute. Let's see how this plays out." She remained hunkered in the dark light of the trees bordering the property. The sloping hill down to the mill property was protected by a tall chain-link fence.

The driver opened up the back door and a man dressed in a long, flowing white kandura stepped out and walked around the back of the car as the passenger continued to point the gun at the man who had come out from inside the building. The driver also had his hand on the gun at his hip.

Clara turned around to look at Marcel with surprise in her eyes and all of the sudden lost her balance, falling backward onto a pile of duff. The sounds of sticks cracking echoed loudly down the draw.

All the men turned and looked up the hill. The driver pulled his gun and started firing in their direction. Clara and Marcel turned to run; their car was just a few paces back down the game trail. Marcel stumbled and moaned, grabbing his arm, then looked down to realize his hand was covered in blood. Adrenaline pulsed through them both. Clara grabbed his bloody hand and dragged him, running back to the car. They jumped in, whipped a U-turn, and sped back in the direction they came from, turning off onto an old narrow dirt road that Clara knew cut through the forest and back to the interstate.

It was definitely the road less traveled. Clara only knew this forest road because there was good climbing back in there; her roommates had taken her a few times. Clara realized she was hyperventilating. She couldn't focus on the road and on Marcel at the same time. She knew she needed to pull over and help him. She wasn't sure how injured he was. He was in shock and kept saying, "I can't believe they shot me."

After about ten minutes of driving, when she was sure they weren't being followed, she steered the car into a pull-off on the narrow dirt road. Marcel moaned and looked very pale. He was starting to sweat.

"I have watched enough crime dramas and hospital

shows to know that needs a tourniquet to stop the bleeding until we can get you to the hospital." She jumped out of the car and ran around to his side, opening the door to have better access to his arm. He leaned out of the car when she opened the door and threw up.

Clara and I moved into the big communal house a few months ago on a six-month lease. The first time I saw the house, I was stunned by its Victorian beauty. An intricate stained-glass panel straddled the doorway, reflecting light off its cobalt and ruby triangles. A turret stood proud on the right side of the house, signaling to the heavens with its pointed patinated top that it had been around for nearly 100 years and planned to be around for at least another century. I couldn't believe I would get to live in a place like this, a real Queen Anne. The brick was deep red, accented by greyish-beige grout lines. Concrete pillars held up the porch. The pitch of the slate roof drew the eye as it looked to slope in an almost asymmetrical concave way.

One of my favorite things in the house was the worn wooden stairs leading up to my room. How many feet had to walk these stairs to create the slight concavity I fill with my feet each time I ascend and descend? I make up stories about the feet that caused the wear. Did they belong to small children? To elders? To a woman who lost her son to war? To a doting husband who lovingly brought soups to his ailing wife. The house held many stories. I was just a passenger for a short time here, only a little over a month left until graduation and then I would say goodbye.

Of my five new housemates, the first I met was Kiki Smith. She was a tiny African American woman with black hair capped with dyed blue tips. She seemed to flutter above the ground and had a tone in her voice that was like hearing a melody every time she spoke. I knew when I met her that we were meant to be friends for life. Her intellect and energy brought out a sense of curiosity in me that was refreshing.

Clara and I had rooms in the old part of the house and shared a Jack-and-Jill style bathroom. Carter Jenkins had the main bedroom in the same part of the house. About 15 years ago, when his parents bought the house, they knocked out the back wall. I guess it was a bit of a fight with the local historical society, but they won and added another three bedrooms, two bathrooms, and a gorgeous gourmet kitchen.

When I first met Carter, I had been a little judgmental. I mean, he was a walking fashion crisis. He wears mandals, and anyone who wears socks with Birkenstocks is just asking for fashion judgment. But he was a gentle spirit and an incredible cook, so I couldn't judge him too harshly, especially when I learned about his story from our roommate, Nik.

Carter's mother passed away from a stroke about ten years ago, and then his father died two years ago, leaving him the house. Given he had a free place to live, he chose the University of Kentucky to get his PhD in

Anthropology. I can only imagine how hard it was—or maybe how comforting it was—for him to live here after they had passed.

The other three roommates, Marcel Harris, Logan Sherp, and Nik Lewis were an interesting collection of personalities. I have been insanely busy trying to close out my senior year of college to graduate summa cum laude, and so I hadn't had a lot of time to get to know them well. They were always on the move, as was I. Nik suggested a few months ago that we share Sunday dinners together, and I am glad he did because that seemed to be about the only time I saw him and Sherp. It was a house full of life and of stories. As much as I yearned to get out of Kentucky, I did love the traditions here, the richness of the seasons, this school, this house, and my friends.

Ping, ping, ping, ping. It was the staccato sound that usually indicated some sort of annoying group text. But it was all Clara. Fifteen missed texts and five missed calls. I don't even read the texts. I told Siri to call Clara. The phone started ringing.

"Mimi Martin, where have you been?! Seriously, can you believe what happened? I mean, he's going to be okay, but that was crazy," Clara spoke so rapidly that she ran out of breath.

"Wait, what happened? I just turned my phone on. I've been at the library all day. I'm just now seeing I missed a bunch of calls from you. I haven't read the texts or anything yet," I said.

"Marcel was shot, Mimi."

I just about dropped the phone. I forget to breathe, covering my mouth in shock as Clara continued.

"We went to poke around that mill and there was some bad shit happening there. I don't know what I thought I would find. It was just using it as a chance to hang out with Marcel. I thought it would be fun. But these guys all had guns and then I fell down and they shot at us and one of the bullets hit Marcel in the arm. It has been the absolute most crazy day of my life. I went to the hospital and then talked to the police. It is like a crazy dream."

I could tell she was still trying to process what had happened.

"Where are you now?"

"Back at the house. We're all here. Are you coming home soon? I really need my best friend right now."

"Yes, of course. I'll be there in ten minutes." I already switched directions and headed back towards the house. I couldn't walk fast enough. I nearly sprinted the last few blocks.

I walked in the door. All the housemates were seated at the big wooden table in the dining room. Marcel's back was to me. I could see his arm bandaged up.

"I hope someone is ordering pizza," he said. "After a day like today, I AM STARVING."

A forced laugh came from most of the group. It was clear no one knew how to react to what happened.

I dropped my backpack on the couch as Clara looked

up and saw me. She ran towards me for a hug. I am not really a hugger. At all. But today, and for Clara, I would have hugged her for eternity. What if that bullet had hit her? I don't know what I would do. She is what keeps me grounded and sane. She is the only one who really knows me.

"Meems, seriously, I'm so glad you are here. Ugh, today has been just the craziest day ever." Clara threw her arms around my neck. She gives real hugs, the ones where she squeezes you for a few extra seconds, the ones I usually try desperately to get out of.

As she released me, I walked over to Marcel. I had to see with my own eyes how he was doing. We were not that close, but he was my roommate.

"Jesus, Marcel, I am so sorry."

"Could've been worse, that's for sure. We went looking for an adventure and found it. I guess we just didn't know what kind of adventure we would find." He was trying to make light of the situation.

"So, what exactly were you all looking for?" asked Nik.

Clara stumbled over her words a bit, "You know, I'm not really sure. I'd been doing some investigative digging after hearing Kiki talk about her research. I'd read that there had been a big shipment of illegal rosewood seized around here a few weeks back. Authorities thought it was destined to be milled in the local area. I knew there was that new mill site, and I thought, well, maybe there is a connection. Lots of rich people want to buy something exotic. I just wanted to see what a place like that looked like. I've never been to a mill or anywhere that works with wood, really. So, I had no idea."

"And you told all of this to the police, I hope?" I asked Clara.

"Yes, of course, I did. The regular police took a statement and then a young Latina woman who said she was a special agent or something, she asked me all kinds of questions." The tears came crashing down Clara's cheeks as she spoke. "Kiki, I did give her your name. I am sorry. I was just so scared and talking and I said you did some research on trees. I hope you don't have to talk to them, too."

"Actually, I would love to talk to them. If I can help solve a rosewood case, that'd be pretty freaking cool, so don't worry about it." Kiki stood up and walked down the two steps into the kitchen, talking over her shoulder. "I think we all could use a beer right about now. Sherp, come give me a hand."

As Sherp stood up to help Kiki, he looked at Marcel. "I am glad you didn't die, man. That would have been a major bummer."

The pizza came, they all had a beer, and then Marcel began his slow exit. Clara stood to walk with him and they disappeared together. Everyone got up, ready to retreat to our own spaces. I noticed Carter looking at me and turned towards him.

"Pretty heavy stuff. I am glad everyone was okay."

"I know. The whole thing is just surreal," I said as I took the last drink of my beer.

"Clara was super worried when she couldn't reach you. It must be really nice to have a friend who cares like that." He looked at me in a strange way. His energy pulled me in. "I think I'm going to go play a few songs upstairs and unwind. Would that bother you? Are you trying to go to bed?"

"Not at all. If it ever bothers me, I'll let you know. I have some calls to make, so you won't be keeping me up."

"Well, if you want to come hang out, you know where to find me." Carter slid down the bench and made his way toward the stairs.

I sat there for a minute trying to process everything that happened. This was like a dream. My best friend almost died or was in a situation where she could have died. My roommate was shot. And I was just oblivious, studying at the library.

As I made my way up the stairs, I heard the guitar through Carter's door. My heart wanted to go in and listen, but my head won.

I dialed my parents' number as I closed the door to my room. I wasn't sure how I was going to share all of this with them, but knew I had to. They are like second parents to Clara. She basically lived at my house when her parents divorced. We've been roommates all through college.

My parents were planning to come up and take us to Keeneland in a few days. I wanted them to know what happened before they saw Clara. I was glad this day was almost over.

The rest of the week was some sort of Salvador Dali painting, with time disappearing and distorted. The weekend arrived in an instant. I studied nonstop for my

LSAT test on Friday. It was a great distraction from everything that was happening.

EXAMINATION

The feeling was uncommon, but not completely unfamiliar. The sweaty palms, the queasy stomach. *I can do this. I have prepared. I have studied endlessly. I was born into this line of thinking.* I half listen as the proctor reviews the rules of the LSAT exam. The buzz in my ears has gotten worse. I looked around the room and noticed that half of us were women, which was nice to see, but of the 30 of us, only a handful had any pigment to their skin. I wondered when the minority would become the majority. And there it was, the bell indicating the test was beginning.

Three hours later, I am a sweaty mess. Time to reward myself with yet another cup of coffee. I texted Clara, begging her to come meet me at The Bean, which was about a half mile from our house, just on the edge of campus. It was a Friday afternoon, so it would probably be packed. She was already in the area and texted back immediately with an *Of course! See you in 15!*

"Clara, it was crazy. I think I was having some sort of anxiety attack in there. *ME!* Anxiety? I am normally the calm, cool, collected one out of the bunch. I can't imagine what it feels like for people who have to live this way all the time."

"Well," said Clara, "the LSAT pretty much determines your future, so I'd say it is as good a reason as any to have an anxiety attack."

We navigated the crowded coffee shop, shimmying sideways between the colorful two-top tables.

"Let's grab our apple table," I said. Clara nodded and made a beeline for our table in the back of the room. We loved this table because we could watch all of the people coming in and out, and it was just the right distance from the stage for their open mic night. We sat and scanned the crowd.

"Ten o'clock, dark shirt, forearm tattoo. I've never seen him here before."

Clara gave me an amused look. It was not often that I put up the guy radar. The corner of her lip curled, and her right eyebrow arched, the way she does when she has something mischievous planned. I stopped her right there.

"Ugh, fine, you are no fun, Meems. You never take any risks. One of these days you need to stop looking and start acting."

I watched Mr. 10 O'clock order his coffee and then turn to leave. He caught my gaze and held it for a moment, dropping his chin as he turned with a smile. It was like time stopped for a moment; I didn't hear anything that Clara said. I was mesmerized.

"Let's head back home. Despite this being my fourth cup of coffee, I am totally exhausted and need to sleep."

Clara was big into music festivals. As we walked home, she caught me up on her plans to head to one in Florida next week. She kept trying to recruit me to go. I love live music but hate big crowds. Plus, I was laser-focused on getting to the finish line with school. I knew Clara was still beating herself up over Marcel getting shot. Maybe I should go just to be there for her?

"You doing ok there, friend? You seem a little off."

"If I had not been so stupid to pretend to investigate, if we had just gone to the gate, like he had suggested, to ask for a tour of the place, he wouldn't have been shot. I mean, what if he can't use his arm again?"

"Clara, you had no idea. Seriously, what are the chances of something like that happening? You cannot blame yourself, at all. Plus, thank God, but he is fine, a few stitches. He said the doctors told him he would make a full recovery, so you can stop beating yourself up."

As we headed up the stairs to the house, Clara hatched a plan. I could see it as she turned to me with her eyes sparkling.

"I think we need to go out this weekend, you know, to celebrate you finally being done with LSATs, and for me to take my mind off all this craziness. We can go to Keeneland tomorrow with your parents and then out to the Club after," Clara said as less of a question and more of a statement.

She was a force to be reckoned with. Her curly blonde-red hair and bright blue eyes, her little nose ring. Her audacious personality. She was the most outgoing and hilarious person. She called herself an "opportunivore" and it did seem that opportunity always came her way. It was kind of like how she always wins free concert tickets from radio shows. Fate was almost always in her favor. Maybe it was the universe paying her back for the hell she went through during her parents' divorce. She has been my best friend since fourth grade. Despite us experiencing life in vastly different ranges of emotions and comfort zones, she gets me in a way no one else does. I liked to observe and analyze life. Clara liked to live it. Somehow, we totally complement each other.

A DAY AT THE RACES

The bell rang, the gates opened, and the mud flew. A surge of well-dressed people lost their inhibitions and screamed at the horses. They screamed like the horses could understand them, but I was pretty sure the only thing the horses heard was their own breath, their heartbeat, and the feeling of the jockey's crop. My parents, Niles and Hae Martin, were visiting for the weekend. Every year, we go to the Spring Meet at Keeneland. It's a tradition that many Kentuckians value. The dull grey of winter has passed, the hills are alive with a vibrant green, and people are ready to get outside and celebrate. We always try to go on opening weekend, usually the first weekend in April, before the pollen and the humidity set in.

I personally enjoy the drive here the most, especially past the big castle. I remembered the first time I saw it in sixth grade, when we started this tradition. I couldn't believe that out here, in the middle of nowhere Kentucky, there was a castle. The rolling green fields always looked so alive; the stone walls lining the road told stories of the Scot-Irish immigrants who settled this area. I am always promised partially burnt brats, courtesy of my father and the tailgating grill that he sets up out of the back of our car. This was always coupled with some sort of history lesson on the area and bad dad jokes—the price we pay for those burnt brats.

Clara joined us as she has for so many of our years. Every year, she seemed to walk away with more money than she came with. This time, she was still a bit frazzled from what had happened with Marcel earlier in the week.

"I won, I won, I won! I should have put down more instead of that measly $6! I hit the trifecta. Look! Red Whiskey, Storm Away, and Bedderuns. In the right order! I wonder what my payout will be," said Clara excitedly.

My mom and I looked at each other and laughed; Clara had won, again. I was genuinely happy to see Clara smiling again. The pony riders found their horses and escorted them off the track as we headed towards the Grandstand for Clara to collect her winnings. It was not quite the Derby—that happens in May in Louisville—but people have been getting increasingly dressed up at Keeneland.

As a young girl, I loved seeing the fancy hats and the smell of the cigars. Now, I am more cynical and wrestle my inner demons about being here. I love coming with my family. I love our tradition. But I really struggle with the historical fact that Black people were enslaved to ride for the sport, not to mention the recent stories that have come out about the cruel treatment of the horses in training. It is all just a little too much for me to really enjoy myself here fully the way I used to.

"How did you pick them this time?" my mom asked Clara as we walked to the window.

"I almost always go by color, unless, of course, one of them just has a totally fab name."

They shared a good chuckle around some of the better horse names they have heard.

"I personally like it when the horses have literary names, but you don't see too many of those. Mostly just versions of whiskey or gemstones these days."

I watched them talk. I loved that they were still so close. Clara had caught my parents up on the awful situation with Marcel over burnt brats in the parking lot. My dad, his usual self, asked all kinds of questions without actually asking them. My mom, stoic and resolved, just listened. Both of them supported her as if she were their daughter. I was glad that my parents had only gently inquired about this round of the LSAT. I didn't want to tell them how poorly I think it went.

My dad stepped away for a call and missed the race and Clara's big win. I saw him walking back through the stone wall cutout that led from the Grandstand to the Paddock. He looked frazzled. He was working a tough case. He retired almost two years ago but has been consulting a bit since then. He sold his portion of his law firm last year, indicating another real step towards actual retirement. He had a noble goal of helping those who cannot help themselves, taking cases all over the world that dealt with human rights violations, human trafficking, and that kind of stuff. Really, his team was prosecuting the worst of humanity. Yet he somehow managed to keep his humanity and his heart intact and still had a lot of humble and authentic love to put into the world, and into our family.

My attention was drawn back to the moment by my mother's voice.

"Let's get a drink before we head out to the Paddock."

I lost sight of my father but assumed he knew where we would be. We always looked at the horses from the terrace just to the left of the stairs.

"Sounds good. A bourbon for me." I smiled at her.

"Me too," chimed in Clara, beaming with pride, "and I'm buying. I just won, like, $150!"

"Keep your money, Clara. You know that when we come here it is our treat." My mom smiled.

My mother's floral print dress flowed behind her as we made our way up the stairs to where the better drinks are sold. Between all of the fancy hats and suits and polo shirts and pointy shoes, I saw a nice mix of the authentic cowboy boots, and the stumbling drunk frat boy who was here for the first time. It really was a coming together of all types at the track.

At the top of the stairs, I was a bit surprised to see a group of what looked to be about a dozen Chinese businessmen. Now that was something you did not see here often! We would regularly get a Saudi prince or contingents of Japanese horse owners here. But a group of Chinese businessmen? That was new. As we waited in line to order our drinks, I realized there were only six of the men talking, and another six men watching. I loved seeing how people from other cultures interacted. Fascinating, and so bizarre sometimes.

We got our drinks and headed back down to the Paddock. Dad was pacing off to the side. I touched his arm to hand him his drink and he almost jumped out of his skin.

"Wow, Dad, what's up? You seem on edge. Did you just lose a huge bet? Scared to tell Mom how much?" I said, trying to make light of the situation.

"Actually, I'm just thinking about this consulting work I am doing right now. Sorry to be off in another world. Looks like I am going to have to head to Vietnam tomorrow. There are some people there who I have a chance to interview, and I'm not sure how long that window of opportunity will last. Do you all want to stay for the last race?" he said with a strained voice.

"Nope. Honestly, as much as I love the tradition of coming here with you and spending this time together, I am kind of over the pomp and charade today." I motioned over to a woman who looked like she stepped out of a catalog wearing some of the highest heels I had ever seen, which were super impractical for walking around this place.

"I'll let Mom and Clara know." I caught Mom's eye and motioned them over.

My dad placed his full bourbon cup in the trash can as we left. We walked out through the red brick arch, past the bronze statues, leaving the noise of the track behind for the quiet of the rolling hills as we headed out to our car.

As my parents pulled up in front of the house, my father seemed to be back to his old self, offering to tell us a joke.

Clara loved my dad's bad jokes.

"What do you do if you are attacked by a gang of clowns?"

"I dunno. What, Dad?

"Go for the juggler!" He cracked up laughing and then stopped. "Oh, sorry, that one might have been too soon?"

Clara smiled at him. "Maybe a bit too soon, but still funny."

"Okay, let's try that again then. Where do dads like me get their bad jokes?"

"Where Dad, where?" I feigned fake exhaustion with the joke.

"The dad-a-base!"

Mom looked at me and rolled her eyes. We all got out of the car and exchanged hugs and goodbyes. I made my dad promise to find a better joke for next time.

The day at the races completely wore me out. I needed a nap if I was going to make it to the Club with Clara tonight. As I began drifting off, I heard the sound of tiny fingernails clawing their way through my walls. The damn squirrels had found their way back in again. Like a rocket, I propelled out of the bed.

"Fuck, fuck, fuck!" I screamed as I stubbed my toe on my desk chair bumbling for the doorknob.

I was totally going to be late to class again. Professor Holmes was going to kill me. I stumbled into the hallway and realized everything was dark. Then I realized that not only was it Friday and I don't have class tomorrow, but it was also 8 p.m., not 8 a.m.

Catching my breath, I made my way to the

bathroom. The door on the other side of the Jack-and-Jill was open and I saw Clara sitting there already dressed and ready. She wore a green tank top and ripped jeans with a chunky sweater and an armful of colorful bracelets.

"The wall squirrels or mice or whatever they are, those tiny fingernailed jerks, are back. We need to tell Carter," I said to her as I shut the bathroom door.

I grabbed a quick shower, brushed my teeth, and put on my uniform of black jeans, black t-shirt, and Doc Martens. It was time to go forget about the LSAT. Forget about Clara and Marcel's close call.

As Clara often reminds me, "Mimi, sometimes you need to just act like the senior you are, with one month left of our undergraduate experience. You have the rest of your life to be serious."

And tonight, she was right.

I had been sneaking into the Club with a fake ID since I was 19 years old. When we were freshmen, Clara had heard about some guy at the local community college who was making fake IDs for $25. So, of course, we bought some. We were so nervous walking up to the door and seeing the bouncer. Correction: *I* was so nervous. I felt some false sense of expectation of how I should behave looming over me. Clara, however, charming as ever, smiled her way to the front of the line, casually chatting up the bouncer. I just nodded and handed over my ID, praying he would not look too hard. He handed it back to me and said, "Have fun, girls." And in we went, no issues at all. Thank God for Clara.

That was the night I realized how much I loved live music, and how much I hated crowds. That was how I found my dark corner. My black fake leather stool. The big wooden beam that I would rest my back on. The sticky drink ledge on the right. It had been my spot ever since. Clara and I had been regulars ever since, usually going at least once a month to see live music. She was always in the front row, dancing her heart out, swinging to the music with whatever guy or gal was next to her. Me, I kept to the back.

The Club, as it was aptly named, had been around forever; a regular stop on the north/south tour route for up-and-coming and semi-popular bands. A guy named Benji Lopez took it over about four years ago, just before I started going there as a freshman. I heard a lot of people comment about how much it changed since he took it over. Better music, less sticky floors, upgraded sound system. And he had really classed up the joint. Renovated most of the inside but left just enough of the essence of the dive bar music scene vibe to keep it feeling real. I felt so bad for him, he did not deserve to die like that, so young . . . The newspaper said it was a car crash. Everyone was shocked when it happened. The Club had been closed for a few weeks afterward. That was a couple of months ago.

As we entered the Club, I thought about Benji and all of these memories. We were legal now, over 21, with real IDs. No more worrying about if tonight would be the night we would be stopped at the door. I sat on my stool and people-watched and chatted with a few people I knew while Clara danced the night away, bringing random shots of tequila, which I accepted with little resistance.

It was probably good that Clara and I left when we did, before I could cause any more trouble. Apparently, I am a total lightweight. I didn't remember much past the first set of music. Except maybe that there was a bit of an incident where I slipped and created a chain event down the bar of people bumping into each other. I recalled there were a lot of drinks spilled. And I absolutely remember this crazy connection I made with the bartender. It felt chemical or magnetic, electric. I am a strong woman who is not afraid to make eye contact, but this was like I was under a spell and could not stop meeting his gaze. Every time I looked his way, he was looking at me. It would have been creepy if he wasn't so attractive. He was the guy from the coffee shop, that forearm tattoo was unmistakable. Mr. 10 O'clock worked at the Club. I couldn't believe I hadn't seen him before; he must be new here.

The phone rang that special tone, the lovely piano piece that indicated my mother was calling. I chewed another sip of the terrible coffee that Clara had made as I tried to recover from last night. My mom always called when my dad left town.

My mom was this tiny, but fierce, yet incredibly calming woman. I guess she lives up to the meaning of her name, Hae, like the ocean. My mom was where I got most of my looks. My maternal grandmother was a beautiful tiny Korean woman, and my grandfather was a tall Black man. Such an unlikely couple back in the day. I got my height from my dad and my grandfather. I stood about 5′6″ with darker, olive-colored skin. My eyes have a hint of my Korean ancestry, dark brown with a slight almond shape. My hair was straight and jet black, like my grandmother's.

"Hey, Mom," I said, trying to sound upbeat and not as terribly hungover as I felt.

"My darling, Mimi, I just wanted to say how much I enjoyed spending time with you and Clara yesterday."

"Yeah, Mom, it was great. Um, can you give me five minutes? I was just about to jump in the shower. I'll call you right back. We can Facetime."

It was one of the longer showers of my life. I stood there for an unknown amount of time trying to wash off the hangover, shame, and anxiety from last night and trying to wrestle with the feelings I had towards Mr. 10 O'clock.

I called my mom back, propping up the phone so I didn't have to hold it. I bundled up in my favorite blanket on my bed and leaned against the wall, which was covered in posters of all of my favorite bands and national parks. Jim James and My Morning Jacket framed my head. The poster I got on our last trip to Zion National Park was in the background.

"Well, there she is." Mom smiled at me. She was sitting at her desk, surrounded by books. "Mimi, my darling girl, tell me everything."

It took me a minute to realize she was asking about

the LSAT. I guess she didn't want to ruin the moment by asking about it too much yesterday. I know she and my dad were trying to not put too much pressure on me.

"Oh. Um, well, I think it went okay."

I am my parent's star student. I am also their *only* student. They always encouraged me to try hard and be my best self. I was the almost, but not quite, class valedictorian. My school in Louisville was a good school, but the surrounding community was, let's just say, not super progressive. They picked a white boy, of course, to represent the best of the school. I might be a little bitter and jaded about this.

"As always, I'm pretty sure I crushed the logic parts. The analytics and reasoning, they were okay, what I expected and studied for." I paused and then added, "I have another shot at bringing up my score if I don't hit the mark this time."

I was an excellent debater from a very young age. From then on, my parents always joked that I would make a fine lawyer. I decided in high school that I wanted to be an environmental lawyer, representing people impacted by climate change and following in my dad's footsteps in helping the world in my own way while also following in my mother's footsteps in terms of being a determined, successful woman. She has always encouraged me to try harder, take risks, to push myself. Her last novel, *The Unraveling*, was her biggest seller yet; and she still makes the time to call and check in on me every few days, despite being on her book tour this season.

"You are a bright girl, Mimi. I know you studied for

this. I am sure you did great."

"Yeah, you are right. I did amazing, I know it."

"Sweetheart, I am guessing you are still pretty shaken up about what happened with Clara and Marcel. Just know we love you and are here for you," I saw the wide-open living room behind her. The piano I grew up at. Her desk surrounded by books. I was nostalgic for home.

"Yeah, Mom, I am fine, really. Just going to hunker down and focus on this last month of school."

She smiled at me, indicating that she understood the conversation was over. We said our goodbyes. My mom blew me a kiss and we hung up. I laid down and pulled the blanket up over my face to try to stop the incessant pounding in my head and realized I really needed a glass of water. I forced myself out of bed and downstairs.

Carter had walked into the kitchen as I filled my glass. "Everything okay?"

"Grand, never better, Mr. Jenkins," I said snidely as I cradled my forehead in one hand, other hand on the glass of water.

"Okay, then. I can take a hint. I hope you feel better soon," he said and walked out.

I watched him. His mandals . . . I still couldn't get over the fact he purposefully wore socks with Birkenstocks, even when it wasn't freezing outside. His long sandy-blonde hair. His soft fingerstyle guitar picking. His demeanor of calm optimism. He was an enigma.

I slowly walked back upstairs and crawled into bed.

LEAVING TOWN

Packing was easy for Niles. He had done it a thousand times at this point. The disheveled lawyer look suited him well. He wore the same rolltop sweater with a t-shirt underneath and blue trousers or jeans. His thick head of dark grey and brown hair was always slightly messy, in an intentional way. This was part of his appeal, what made him so approachable. He could fit in with anyone on the street.

He threw a few changes of clothes, his toothbrush, and an electric razor into his overnight bag. He gathered up his notepads, his recorder, his extra battery charging pack for his phone, his burner phone, and his laptop and placed them in his worn leather briefcase. The briefcase was a gift from Hae, from nearly 25 years ago on their first trip to Florence together. She was working on a novel and needed to conduct some research, so they went on a "vacation" to study the town. The briefcase had held up well after all these years, a testament to the quality of Italian leather products. He was packed and ready.

He walked down the stairs to find Hae at her desk writing. She looked up at him, a smile on her face as she took a sip of her tea.

"It was really lovely seeing the girls at Keeneland, wasn't it?" Hae stood up and walked toward him.

Niles was busy in the bureau looking for something

and turned towards her.

"I am glad we have kept up with our tradition. I only hope that after Mimi goes away to law school or wherever she ends up, that we can keep doing it."

"How early is your flight?" Hae asked.

"I need to leave here by 4 a.m. It is a nightmare of a schedule. First to Chicago, then a stopover in Dubai, then onto Ho Chi Minh. Since I have to stop in Dubai anyhow, I am going to pay a visit to Mahreen. I haven't seen him in ages. I should be headed home by Saturday, at the latest. Hopefully back in time to see you off before you leave for Denver for your talk on Sunday."

Niles leaned in and kissed Hae's smooth, flawless cheek. He loved the way her skin always glowed. She had barely aged since he met her.

"I am sure you will be up late writing, as usual. So, this is goodnight and I love you. Four a.m. is going to come quick. See you Saturday night," he said as he wrapped his arms around her waist. She leaned into his chest with the tip of her head. She was at least four inches shorter than he was, which made it easy for her to tuck up under his arm.

"Yes, and please call when you get there. Safe travels, my love." Hae rose onto the balls of her feet for a goodnight kiss and then collected her teacup off the bureau and made her way back to her desk.

The morning came very early, but no traffic and flights were on time, putting Niles into Dubai around 8 p.m.

A driver met Niles at the airport, finding him by holding a tiny electronic sign with his last name, "MARTIN," in large letters. He followed the driver out to the black Audi. He opened the trunk and Niles declined, walking towards the back passenger door. The driver, dressed in all black, casual attire, opened the door. Niles tossed his bags in the backseat and climbed in. Niles didn't know the destination but knew when the car stopped that Mahreen would be waiting for him. He was right.

"From the beautiful rolling hills of Kentucky to the beautiful deserts of Emirates, you are a lucky man, my friend," Mahreen said as Niles exited the car into a deep handshake turned hug.

"It has been too long, my friend. Your country is just as beautiful as I remember," Niles responded, stepping back and scanning his surroundings.

"Welcome to my home," Mahreen said as he led Niles inside. "Please, you can leave your bags there, and when we are finished, my driver can take you to your hotel." He motioned to an area next to a table that was laden with marble statue heads.

"What is it that was so important you had to fly all the way across the world to talk to me about?" Mahreen inquired as he motioned to his servant to pour them two drinks then turned halfway around to smile at Niles. "Ice or no ice?"

"Mahreen, no one puts ice in scotch," Niles said,

laughing.

"Ah, you are correct, and still a good man with good taste I see," Mahreen countered.

"Let us go out to the terrace and talk," he said as he handed Niles the small glass of scotch and raised his glass to toast.

"There is a case I am working on," Niles began.

"A case? What happened to retirement, my friend? I will tell you; the retired life is very good." He motioned his hand around his terrace.

"Ah, yes, I can see it is treating you especially well. And myself, as well. This is a special situation, just one I am consulting on in retirement. I have been doing some research and talking to many sources. In my mind, I have found some connections, but I cannot find them in reality. I wanted to get your opinion on such a matter." Niles sipped the scotch, sat back in his chair, and paused.

"It depends on the matter, whether or not I can help you, but please continue," Mahreen offered.

"So, I cannot find a clear connection, but all paths of my research are leading me to believe that somehow the trade of illegal exotic wood and the trade of people are tied together. Also, somehow these are linked to someone in the thoroughbred racing industry," Niles said carefully and quietly.

"And you come to me because of my Saudi connections in the horse racing industry, is that it?" Mahreen posited and Niles nodded.

"I cannot say I have heard of any connections, but I can be sure to listen carefully to what I am hearing in these circles," Mahreen whispered and then he spoke in a normal tone. "What I do know is that the trade of illegally harvested exotic wood is on the rise. In my country, we have very few forests, maybe less than 5% of the entire country. There is high demand here for quality wood, as is the case in much of the world. Climate change is upon us and there is a fear that these precious trees will disappear completely, driving up prices and feeding the black market."

Niles tipped up his glass of scotch and then leaned forward onto his elbows. "I know, and it also seems that the human trafficking cases uncovered lately have women from these very areas where we think the illegal harvesting is mostly occurring. I am trying to find the link to the horse industry. Maybe there is none. Maybe it is just coincidence or a bad lead."

Niles sighed and sat back in his chair, looking at the white lit walls surrounding the terrace. His mind lingered on some tough cases from years ago where he had helped prosecute about a half-dozen human traffickers. Hae had helped him find recovery homes for some of the women, to help them re-enter society after what they had been through. Many of them were forced into prostitution, starting as teenagers. Some were taken as they tried to cross the border. Some were sold a false narrative of going to work as a housekeeper for a rich couple in New York. How they got there did not matter. Niles felt such disgust and disdain for the people who were involved in these trafficking rings.

This new connection to the timber industry was interesting. With enough proactive effort and education, it is possible that nonprofits could help make people aware of the danger so they could prepare. The link to the horse industry had to be for the shipment or transit, he couldn't quite figure it out. He brought himself back to the present as Mahreen was telling him about his children.

They spent time with small talk of days past and enjoyed the remainder of the scotch.

"Our times together at Georgetown those many moons ago, who would have ever thought we would be living these lives now?" Mahreen said as he motioned to his surroundings.

"I thank you for your time and your hospitality, Mahreen. It is always a pleasure to see you," Niles said as he stood to leave.

"My friend, must you leave so soon? Next time, you must come during the daytime so I can show you my gardens." Mahreen rose to walk Niles out. The driver was waiting to take Niles to the hotel.

"Ah, yes, I am happy to see your business endeavors have treated you well, being able to end up here in Al Bahari. There is certainly still money to be had in oil." Niles stuck out his hand, looked Mahreen directly in the eye, and nodded.

He turned to the car as the driver opened his door and placed his bags next to him on the backseat.

Driving away from the house, Niles noticed the view of the lit towers of downtown Dubai. They were quite breathtaking. As a man who loved the country and fresh air, there was a certain appeal to this city that got him every time.

At the hotel, he looked out at the turquoise illuminated pools and the downlit beaches, noticing a Ferris wheel just down the beach. He pulled his shades closed and turned on the TV news. He sat with his notepad and jotted down a few thoughts from the day.

He awoke in the morning with the TV still tuned to Al Jazeera. Opening his shades on the far side of the room, he looked out at the ocean. Here he had come halfway around the world for an hour with an old friend, and no closer to the information he was seeking. Hopefully, Vietnam would yield more results.

The airport in Dubai was a fascinating place. A mixture of the modesty of Emirates combined with traveler cultures from around the globe. Walking the corridor, you see ghutras, kanduras, and head scarves alongside saris, business suits, and loungewear. It was a massive airport.

Niles walked for nearly 20 minutes to his gate, finally arriving as they were checking passports. He had his documents checked and stepped back to scan his lunch options, ultimately settling on a noodle shop for a quick bite before boarding. It was one of his golden rules to never fly on an empty stomach, especially for anything over four hours. Vietnam was a little over seven hours away.

The Emirates flight was smooth into Ho Chi Minh City International Airport. He was met by a finely dressed woman, who, upon greeting, he realized was American. He learned that she works on contract with the State Department. She was quiet and not much exchange happened on the drive to the hotel, which was fine by Niles as the jet lag was beginning to set in.

As they pulled to the curb of the hotel, she informed him she would be back tomorrow at 9 a.m. to pick him up. He glanced out of the window of the car to make sure no mopeds or bicycles were speeding down the road towards him. The onslaught of the sound here was always so surprising. It was the sound of life, of machinery, of development, of many, many humans all trying to navigate the same space. It was harmonious and chaotic all at the same time.

Niles stepped into the Orchids Saigon Hotel, an opulent place. He looked up and noticed the chandeliers that consumed a massive amount of ceiling space casting light fragments onto the marble floor.

"Welcome to Orchids Saigon. Checking in, sir?" said the neatly dressed Vietnamese woman behind the counter.

"Hello. Yes, thank you. Martin, Niles Martin." He smiled at the woman. "If possible, somewhere away from the elevators and any ice machines and close to the stairs, please," he said politely.

He scanned the lobby, noting exits and assessing the overall situation. Many years of training made this type of assessment second nature. After several minutes, the hotel staff finished her process and handed him his key.

"The elevators are over there. Fifth floor. Enjoy your stay. Room service ends at 11."

ONE CHILD POLICY

I must have fallen asleep after talking to my mom yesterday afternoon and slept the entire day and night away, which was completely abnormal for me. I had not set my alarm. It was already 7:30 a.m. and my first class was at 8. Crap. My right arm was completely asleep. I dressed, combed my hair, grabbed my satchel in record time and ran to the steps when I looked up and saw Carter coming out of his room. I almost took him out.

"Hey, looks like you are in a hurry. Need a ride?" he asked.

"Um, no, but thanks, I'm okay," I said, running down the stairs and out the door. I sprinted the five blocks to the edge of campus, across the lawn in front of the Administration Building, and up the stairs to my political science class.

Those stairs, every time. You would think I sit on the couch all day for how out of breath I was. I really need an A in this class to keep my GPA high for applying to law schools. Plus, today we were going to be talking about China's population control policies and I did not want to miss this. I needed to impress our new professor, Dr. Pile. He just took over this class from our old professor who had to step away for some family issues. I really like Dr. Pile's approach. It was only his second class with us, but last time he had us divide up and debate a topic. I thrive in that environment. Was it weird to actually enjoy school like this?

"It is absolutely a human rights violation to force women into how many children they can or cannot have," the girl in the pink tank top said as I slunk into the seat in the back of the room and sat down.

"It is not. There are finite resources and China was just trying to make sure that they could provide for every one of their citizens," said the frat boy in the front with his perfectly coifed hair, khaki pants, and polo shirt.

"You cannot seriously believe that." I found myself saying. "It would be one thing if you stated that the limitations were mostly due to the high cost of living and inability to provide for more than one child. That could be an argument I would engage in. But overall resource availability, nope, not buying it. It is obvious this failed policy has resulted in an aging population, limited the ability of many rural areas to have economic growth. There are way more men than women, which perpetuates the fertility and gender skewing crisis in the country. And do we really want one of the largest countries in the world being dominated by masculinity? We need the feminine element to balance. The yin and yang. I think that policy will have horrible ripple effects for China, and for the world for at least the next several generations." I sat back, realizing how fired up I felt. I am passionate about certain things. When I get going, I really get going.

The debate continued and, looking up, I realized Dr. Pile was closing his book, which always signaled the end of class. The normally completely focused and engaged me was somewhere else the rest of class, I guess. My mind was on my pending LSAT score and still lingered just a tiny bit on Mr. 10 O'clock.

"Mimi, can you come here for a moment, please?" I heard Dr. Pile say in his deep voice, which I have always thought would make for great radio. He looked like a textbook poli sci professor: corduroy blazer with patches on the elbows, deep brown skin highlighted by his mostly white beard.

"I really appreciated your thoughts today," he said. "What is it you plan to do after you graduate?"

"Law school, but first a gap year. I graduate this semester, and am looking for job clerking or interning, and might travel a little, and hope to go to Harvard Law next year," I said and looked down at my feet, realizing there was something on my Doc Martens. It was unlike me to be this distracted. Plus, it was a little strange for a new professor like this to take such an interest in my life.

"Very good. I think you have a lot of potential, no matter what you choose to do next. It was nice to chat with you," he said, smiling.

I gave him a nod and made my way up the stairs to the back of the auditorium-style classroom. I got this weird, unsettling feeling that he watched me exit. I looked over my shoulder and he was standing there, arms crossed, laser-focused on my departure, and flashed me a broad smile. I quickly stepped out into the sea of students swarming between classes.

VIETNAM

Outside at 9 the next morning, the driver stood waiting for Niles. They drove past the old US Embassy that was decommissioned in 1975. The current offices of the consulate are much smaller and just a few blocks from the hotel, which Niles noted. He could almost walk there faster than drive.

The sidewalks were lined with trees on one side and the tall protective walls of businesses and residences on the other. Most of the buildings were a beige color, some colorfully accented by signage or decorations. It was not very busy in this part of the city.

Once at the consulate, Niles' credentials were checked and passed through the scanners. Then he was escorted down the hallway. Halfway through, he was intercepted by Clyde Spees, the Agricultural Attaché he was meeting with this morning. They walked together to the conference room, which was a relatively dull and muted space with a large wooden table and black cloth chairs.

"It is nice to finally meet you, Mr. Martin," Clyde said as he moved to the other side of the table and pulled out a chair. "So, you are here to learn more about the world of international timber trade, is that correct?"

"That is correct. I am on contract for a non-profit organization that is looking to combat the illegal trade of rare and exotic woods. I understand you have quite the specialty in this area. I was hoping you could fill me in on what is driving this apparent increase in illegal logging," Niles said casually.

"A non-profit? I was told you were a former Justice Department lawyer?" Clyde said. His tone was either slightly amused or annoyed, Niles was not sure which.

"Well, I am that, too. Retired. Now just working to help make the world a better place however I can." With a smile, Niles placed the recording device on the table and asked, "Do you mind if I record our conversation?"

"Sure."

"Can you start by telling me a little about yourself and how you got to be such an expert in illegal logging," said Niles.

After about 45 minutes of interviewing, Niles realized he was no closer to the answers he was looking for than when he entered. Clyde was only providing information already readily available. Nothing insider; nothing new. Vietnam only had small-scale illegal logging issues. Vietnam was ramping up exports but was hindered by numerous laws. Vietnam banned cutting of its natural forests years ago and relies on plantations of rubberwood and acacia, but most of this wood is too small for furniture and other products, so are sold as wood chips. Because Vietnamese companies cannot cut natural forests and the plantation trees are too small, manufacturers in Vietnam import wood from all over the world and, in many cases, imports from areas with illegal logging.

The US has been working closely with Vietnam for years to develop a certification program for imported wood products to ensure they are legally harvested. While this effort has gained wide government support and traction, the demand for black market logs still exists. The Convention on the International Trade of Endangered Species (CITES) has been adding new species of trees to its list every year, increasing the number of trees that are illegally traded. Some of the most critically endangered tree species are even more vulnerable to extinction now due to the combination of climate change.

"Which woods are drawing the highest prices right now on the black market around here?" Niles asked.

Clyde left for a moment to retrieve some samples of the seized wood.

"Rosewood is the biggest. In China, a cubic meter can go for $50K or more. Ebony is another one we see trafficked a lot around here."

He continued, "We've worked closely with governments across Asia to really try to cooperate to reduce illegal logging across the continent. The real issue is Brazil. A lot of issues with tracking down there."

"Yes, I'm aware they have quite the challenge," Niles said before reaching over and clicking off the recorder. "Thank you, Clyde. I really appreciate your time and your passion for what you are doing. Before I go, do you have anyone else you could recommend I talk to while I am visiting?"

"There is an NGO here who works on land tenure issues. That is closely linked to the issue of illegal logging. I have been tracking a bit of their work. They might have some different perspectives to share with you," Clyde pulled out his phone, wrote a name and phone number on a sticky note and handed it to Niles.

"I'll call my contacts over there today and see if I can set up a time for you to meet with someone tomorrow," Clyde said as he slid his chair out from the table.

"Wonderful. I look forward to it. Thanks, Clyde. I hope our paths cross again and best of luck. Let me know if I can ever help you."

Clyde escorted Niles down the hallway and towards the exit.

Niles stepped back out onto the street and began walking back towards the hotel. He stopped by a street vendor for a bahn mi en route and half-noticed a man across the street who seemed a bit out of place.

He walked quickly back to the hotel, casually crossing the street several times, giving him a chance to see if he was being followed. He didn't see the man from the bahn mi stand any longer.

The Orchid Hotel's lobby brimmed with life. There must have been a large conference or event taking place; every cocktail table and booth in the hotel bar upon entry was occupied by business casual travelers, their chatter echoing off the tall marble walls. There was a line to check-in. Niles worked his way through the people and to the elevator. In his room, he looked around to make sure everything looked normal. Nothing appeared suspicious. He phished the burner phone out of his bag. Niles was on edge. He called his long-time colleague and friend, David, whom he had just reconnected with a month or so ago.

"Morning, David. Just checking in. Hope I didn't

wake you." It was around 5 in the morning for David.

"Morning, Niles. You know I am always up before the sun. Vietnam providing any leads?"

"I am not getting any hits so far, but I have a meeting tomorrow with a local non-profit that seems promising. Also, any word on the shooting at the mill off Big Ridge Road last week?

"Agent Rodriguez has it covered. We have the whole team in DC working on this, as well," said David.

"Good."

"Niles, I have some details to read you in on when you return from Vietnam. Just know we are keeping a close eye on your daughter and her friends. As far as we can tell, they aren't in any danger. They were just in the wrong place at the wrong time."

"Thanks, David. It makes me feel better knowing you all are keeping an eye out."

David and Niles had worked on several cases together over the years. First, they were at the Justice Department in the '90s, both eager, young lawyers. Eventually, through a series of cases and time, they both moved to intelligence work for the FBI, especially after 9/11.

Niles worked at the FBI until 2008 when he decided it was time for a change. He moved his family to Kentucky and left the fast-paced life of Washington DC behind. There was a different colleague Niles had worked with at the Justice Department who had also left and decided to open up his own law firm in Louisville. Niles decided to join him; he was ready for some semblance of a normal life, especially since his daughter was about to enter middle school. He looked forward to picking his own cases and working at his own pace to make the world a better place. Niles had always wanted to work with the exploited and underserved. This move was his chance to do just that and to prosecute the worst of humanity.

Niles would occasionally be asked to consult on a case for the FBI, so he made sure to keep up with his training. He went to the gym almost every morning and practiced regularly at the shooting range.

"All right then, let's connect when I get back." He flipped the phone shut and set it on the table next to him.

It was time to see some of the city. Niles packed up his satchel and headed out the door.

It was late afternoon, and the air quality was awful. Ho Chi Minh is known to have some of the worst air pollution in the world, but today seemed especially bad. It could have been the stifling heat; it was over 90 degrees today. Ho Chi Minh is the more modern name for the ancient city of Saigon, renamed after the war and reunification of the two parts of the city in 1976.

Walking the streets, Niles noticed the abundance of French architecture left over from the colonialization of the late nineteenth century. He walked down Nguyen Hue Street, through the promenade, and turned towards the Saigon River. He watched the boats go by and life move along the streets.

No stranger to the thieves in the area, Niles secured his bag in front of him as he went. He kept a careful eye around him. Maybe he was just overly suspicious earlier. He didn't notice anyone following him or anything out of place.

Niles settled into a small café near the ferry port. He pulled out his tablet and did some research on the non-profit he would be visiting tomorrow, making notes about what he wanted to ask the contact. Clyde had mentioned that the woman was a bit apprehensive and that Niles should approach it gently. That's his style anyhow: create trust first, find common ground, then move into the questions.

It was about 5 in the evening and the sun was setting. Niles packed up his things and paid his bill then wandered back onto the street. He walked past the Opera House and the Notre Dame Cathedral of Saigon. Both incredible pieces of architecture. He decided to step inside the church.

He always loved visiting the ancient churches of the world. Mimi shared this love with him when she was young. He and Hae had taken her on a few trips to Europe. She was always fascinated by the stained-glass windows. He remembered how her smile would light up an entire room when pipe organ played or the chorus sang.

He slipped into the pew and looked around. This particular church looked newer, perhaps built in the late 1800s. It lacked some of the appeal of temples in this part of the world, but still had soaring bell towers, the arches in the French Gothic style, and notable bright red bricks.

The door opened and shut behind him. A man entered and walked around to the left, lit a candle then walked up the left aisle and slid into a pew a few rows behind Niles. Niles admired the stained glass in the dome above the altar for few minutes before standing to exit.

Across from the main exit was a beautiful lush, green garden with a statue of Mary fixed in its center. Niles crossed the road quickly to the garden and the far side of Mary, looking back at the doors of the church. The man did not exit. *Another false alarm,* thought Niles. He walked quickly back to his hotel.

As promised, the driver was in front of Niles' hotel at 9 a.m. sharp. Niles got in and they wove through traffic to the outskirts of the city where the non-profit, AgriForest, was located.

The office was nondescript, a bland beige building with a simple sign on the front. It stood connected with other businesses in the same frame, along a generally quiet street to the northwest of downtown Hi Chi Minh. Nguyễn Thí was waiting at her tiny desk, hidden behind her computer screen when Niles walked in. She stood and bowed slightly and nodded upon his entrance.

Niles bowed to her, "I'm Niles Martin, I believe Clyde told you about me."

She acknowledged him with a nod. "Please sit, Mr. Martin. I am the only one here today, the rest of the staff are gone on site visits," she pointed to the chair across from her desk then turned off her monitor and leaned onto her elbows on the desk.

"Thank you for agreeing to meet with me." Niles settled into the chair across from her. "I believe Clyde shared this with you already, but I am currently working for a non-profit in the US. One of our focus areas is an effort to stop illegal trafficking, mostly of women and children, but in some cases of other illicit goods. Right now, I am doing research to better understand how illegal logging works."

Niles paused for a moment and pulled out his recorder. "Do you mind if I record our conversation today?"

"These things are better not recorded," she said simply. Niles slipped the recorder back into his bag and pulled out his steno pad.

"I understand you are an expert in this area. I have read about your efforts and how you work tirelessly on behalf of your people to secure land rights for them. I am impressed with the work you have done. I would like to learn more about these efforts and the work you do here."

Thí proceeded to describe the complicated issue of land tenure in Vietnam. The country is less than half forested, and the forests are classified as protected, special use or for production. Thí mainly works with common property or lands managed by small communities, but also in securing private property rights for landowners. Land tenure policies started to shift in the 1990s, from state-seized and operated lands to allowing individuals property rights. She had been working in this area since then. The goal of this forest devolution movement, or giving land back to the people, as she described it, was to alleviate the extreme poverty seen in many parts of Vietnam.

"There have been many complications over time with land seizures, lack of transparency, corruption. In our small communities that manage natural rich and diverse forests, they do not have the resources to protect those forests from the predators of the night," said Thí.

"Can you tell me more about these 'predators of the night?'" Niles glanced up from his notepad and noticed that Thí seemed uneasy.

"The thieves come in at night. They cut down the special trees that we are trying to preserve and take them away. They forge the paperwork to say they were legally harvested, and then they sell them. Or if they cannot get the paperwork from the corrupt state officials, they take them to mills under the cover of night and process them. It is a problem across Asia, these black-market thieves operate in areas where they know there is no force that can stop them," she paused. "Our government has come a long way trying to stop this. There are policies in place. But you cannot stop corruption."

She paused for a moment as if trying to find the right words.

"In other places, farmers who have worked the land for generations have been given rights, but only for 20 years. When that time is up, their land is seized. We are working to secure longer-term legal rights, to work around the state-managed system, to hopefully preserve and conserve these important forests for our children's children and for the livelihoods of many Vietnamese," she sat back in her chair. "Do you have any other questions?"

"Yes, actually, I do. Is it possible that in these impoverished rural areas you speak of, have you heard of other things being taken besides the trees? People, perhaps?"

Thí's face darkened. She lowered her voice and stood, walking around the desk to where Niles sat.

"These are the terrible things that happen with the predators of the night. While some steal our trees, others steal our people. It can be violent. It is a fear that we all have. The government has shown weak attempts to investigate, but still today, many women and children remain missing. It is a great tragedy. And it is one we do not speak of too loudly. Too many ears listening," she whispered and turned to walk back to her desk.

"Do you know who is doing this? Is it random? Organized?"

"I think you know the answer to your own question, Mr. Martin. I think we are done here today."

"One more question, if you have the time. Is there a certain type of tree that is most commonly stolen from these areas? And would you by chance have any examples of it to show me?" Niles asked.

Thí reached across her desk for a small carved elephant. "This is one example, Thai rosewood, or *Dalbergia cochinchinensis*, from the small community of Ngọc Thanh in the Vinh Phuc Province. I keep it as a reminder of what I am fighting for. But remember, here, in our country, while we do have some illegal logging, we are more of importers and launderers of the illegal roundwood and veneer coming from all over Asia. Don't ask too many questions. Be careful who you talk to, Mr. Martin. These are very sensitive topics in our country," she stood, indicating their time was up.

The driver was still waiting outside. Niles failed to notice the car that pulled in behind them as they were leaving.

Back at the hotel, Niles gathered his things in preparation for his flight. He decided to give himself a little extra time at the airport today. The driver remained downstairs waiting for him. His flight was supposed to depart a little after 7 p.m., 30 hours with two stops, getting him back into Louisville at around 3:30 p.m. on Saturday.

He asked the driver if she would mind stopping by a café on the way out. It was nearly 2 p.m. and he had not eaten since breakfast. His phone rang while he ate his Pho at the café. He could see his driver standing next to the car across the street. It was Clyde.

"Martin here," Niles said.

"Niles, did you go see the AgriForest people today?" Clyde sounded incredibly nervous.

"Indeed. Ms. Nguyễn was most helpful, actually. I was impressed with her work—"

"She is dead, Niles. One of her staff came back to the office around lunchtime and found her with a bullet through her head. Be careful, Niles. Be very careful." Clyde hung up without another word.

Niles set his spoon down. He could not fathom what

he just heard. The driver looked over at him as he stood from his chair, studying him carefully. Every inch of Niles went on high alert. He quickly scanned the café. Nothing seemed out of the ordinary. He walked across the sidewalk towards the car.

"We need to go," he said. The driver understood something wasn't right. Her hand instinctively went to her hip where Niles noticed her gun for the first time. She scanned the café and street before they hurried to the car. The driver walked the perimeter of the car and looked underneath it before opening the door. She nodded at Niles and they got in. They took a circuitous route to the airport.

At the airport, Niles thanked the driver and exited quickly into the sea of travelers. He was on edge even after he boarded the plane, his mind racing about what Clyde had shared. Thí had been killed. Niles had been in this business long enough to know that he was on the trail to something, and that someone had been following him. He barely slept during the flights.

Niles was on autopilot during the drive home from the Louisville airport. He didn't have any concept of the time that had passed from getting into his car to pulling into his garage. Sleep-deprived and on high alert, he entered the mudroom that led to the kitchen.

There, Hae was making tea. She smiled warmly. He felt much of his tension fall away. After nearly 25 years of marriage, she still had this effect on him. She was wrapped in her long flowery silk robe.

"Welcome home, love." She walked over, rose onto her tiptoes, and kissed his cheek. "Was it a good trip?"

"You are more beautiful than the full moon you were born under," Niles said as he wrapped Hae in his arms. "I love you."

He paused in that moment and, in her arms, exhaustion hit him. "God, it is good to be home. I am going to head up and grab a shower." He showered, changed, and headed straight to bed.

DOMINO

I needed coffee if I was going to make it through the rest of the day. So, I walked the well-worn dirt path between the business school buildings, noticing for the first time the gorgeous flower beds full of spring beauties and tulips along the way. My favorite tree on the whole campus was here, too. I learned last year that it was not just any magnolia tree, it was a "sweet bay" magnolia. Its buds were just about to burst.

The line at The Bean was almost out the door, which was not uncommon at this time of day. I was glad that my next class was a few hours away. Nearly every person in front of me was looking down at their phones. This drives me crazy. Maybe I'm an anomaly, but I would much rather people-watch and observe the world than stare at a tiny screen. I scanned the line and told myself little stories about each person. Clara and I always enjoyed doing this at The Bean. We made up stories about the people who came in and what they do, where they come from. It was a great way to pass the time and a great diversion from my never-ending reading of real-world politics in the *Economist, Atlantic Monthly,* and various newspapers. It was nice to get lost in fake reality every now and again.

"So, Domino, are you recovered yet?" *Domino?* My cheeks flushed and my heart raced. I pretended that I didn't hear that gravely sexy voice behind me. And then he, Mr. 10 O'clock, was right here next to me, and I broke into a full sweat.

"It looked like you and your friend were having a really great time the other night. What were you celebrating again?" he asked.

I laughed, which was my best attempt to be calm and collected. It was such an awkward laugh . . . "Not sure if I was celebrating or not, but I had just taken the LSAT and was blowing off some steam. I guess you heard about the magical barstool that just happened to move out from underneath me?" I quipped and looked at the ground, digging the toe of my shoe into the ground. Memories of that night had come back to me the next day.

"Oh, yeah, bar dominos. I found it highly entertaining. It's been a while since I've seen something like that. That guy in the khakis, he is the assistant football coach and was not happy about having to drive home smelling like a college kegger. Those barstools look nice but have a sneaky way of moving around." He had a half grin on his face.

I sensed a slight accent in his voice that I could not place. I raised my eyes to meet his gaze and gave a half smile.

"So, Domino, do you have a name?"

"Mimi, but my friends call me Meems. And you are?" He can't forever be Mr. 10 O'clock.

"Damian Lopez. Just moved to town from Miami. So far, this is the only place I have found with decent coffee. In Miami, there is a coffee shop on every corner."

"Yeah, I wouldn't say a small town like this has near

the coffee culture as Miami. But The Bean is solid."

"Well, Mimi, or Meems . . ." and he paused as if he was digesting my nickname. "Do you have a last name?"

"I do, but I am going to need to know you a little better before I share that kind of personal information."

Oh my god, did I just say that? Did I just basically ask this ridiculously hot guy out? What the hell is wrong with me? I'm not casually charismatic.

"It's a date then. Want to meet me at my place?" he stammered for a second then corrected himself. "Not my apartment. That would be too forward, wouldn't it? I should rephrase . . . Would you like to meet me at the Club tonight?"

"The Club is your place?" I said incredulously.

"It is now. That is a story that I will only share if I get to know you a little better." He winked at me. I gave my head a little shake, letting my hair fall over my face to hide my reddening cheeks and the smile I couldn't suppress.

I was suddenly at the front of the line and the barista asked for our orders. I was so distracted I could hardly hear her. I ordered myself an iced coffee to go.

"I've got this. It's on me. I'll have a cappuccino," Damian said.

I couldn't help but do a full belt-to-chin scan of him, somehow managing to stop myself before looking lower. Or higher. I refused to let myself get sucked into those stunning dark brown eyes.

"Thanks. I owe you one. Um, I can't make it tonight, but I'm sure I'll see you at the Club soon. I'm there a lot in the back-right corner by the post."

"I know," he said as I walked away. I briefly glanced back over my shoulder to see him watching me.

Why does everyone keep watching me walk away?

As much as I dreaded moving into a big communal living situation, I now feel like I have found a home here. Even if some of my roomies smoke way too much weed and party a little too much, everyone worked hard and, honestly, were some of the smartest people I had ever met. Clara had somehow convinced me in my drunken stupor last Saturday to join them on this music festival adventure. I couldn't believe that I actually agreed.

The group rented an RV for the trip. It was a pretty smooth ride, which was great considering the 30-hour round trip. It was a Thursday and the bands would be performing all weekend. Skipping class to do something like this was very out of character for me, so I may or may not have had a bit of anxiety around this whole thing. Clara reminded me again, for the millionth time, that I needed to live a little. So, here I was, living.

Carter drove. Marcel sat in the passenger seat. The rest of us piled on the couches in the back. We were about an hour in, headed south. The energy in the RV was high, everyone brimming with excitement for the weekend.

"So, Kiki, I read this amazing book last year about

how trees can talk to each other. I know you have a PhD on trees. Do you think that is real, that they can talk and are all connected?" I asked.

"Ah, I bet you read Simard's work. That's some pretty interesting stuff, but not at all my field of expertise. Her science seems pretty legit, so I'd say it is quite possible. Kind of crazy to actually think those trees out there are all having a conversation just like we are." Kiki looked out the window. "With forensics, I mostly study the dead trees. Simard looks at the living ones."

"What made you pick tree forensics? It just seems so niche. Before I met you, I didn't even know that was a real thing."

"To be honest, it was kind of fate and kind of a bit of trial and error. I knew I loved trees and I knew I loved science. And I knew I loved watching CSI and always thought it would be cool to do something like that, like crime scene forensics. So, when I found out there was something I could do that combined all three, I was all in."

"Are there a lot of jobs out there for that? Or do you think you will become a professor or something like that?"

"Oh, yeah, there are jobs. The global trade of illegal wood is huge. Lives are lost, entire ecosystems devastated, communities uprooted." Kiki was suddenly in her zone. She pivoted in her seat and leaned toward me. "I'm developing new chemical and genetic techniques to trace wood origins. I'm also working to improve genetic indicators used to trace sustainably harvested wood from around the world so that places like Home Depot know that when they stamp something in their store as environmentally friendly, they can prove that it comes from a legit source and isn't like contributing to the deforestation of the Amazon and stuff like that," she said with a big smile.

"That's crazy. I can't believe I haven't heard about it on the news before."

"Oh, it is a huge thing. There is a whole underground market built around this. It is like the drug trade. Thieves come in under the dark of night, paperwork is forged. It is a special kind of evil that these criminals subscribe to. Think elephant poachers and ivory. It is that kind of evil." Her face twisted with disgust. "Who could kill an elephant, one of the most intelligent animals in the world, just for their tusk? We are getting a lot better at tracking the origins of the wood with genetic proof. When we can find the source location then the good guys can start the hunt for the poachers and shut them down for good. I just do the science part; other people have the really hard job of catching the bad guys." The passion in her voice rang through the whole RV.

"Don't get her started," said Sherp. His red hair was jumbled and his torn Phish t-shirt announced that he really liked to smoke pot. "She gets really fired up about this."

"It's fascinating," Clara leaned in, glaring at Sherp. "I think it's really admirable. And you're a great storyteller. I felt your passion. As a journalism major, that is what we look for. I cannot wait to interview you more for my story!"

Marcel, still wearing a sling and bandage from where the bullet grazed the outside of his arm, leaned over from the passenger seat and looked at Clara. "I might be done being your companion for getting the rest of that story." He winked and turned back around.

Marcel Harris comes off as slightly devious and a hopeless romantic. He and Clara had been in some classes before we moved into the house together. That was how we found the place; Marcel had been living there for the last year. He and Clara had been hanging out together a lot the last few months since they were both journalism majors and had a lot of classes together. He planned to travel the world and cover the really hard stuff, notably in places where most people don't want to go but still want to know about.

Nik pulled out a deck of cards. Sherp took a long hit off of a glass bowl filled with weed and passed it towards me.

"Oh no, not my thing," I said and waved my hand at him. I didn't want anything like that jeopardizing my future of getting into law school.

"So, what is your thing, Meems?" Nik asked me.

"Oh, um, you know, trying to save the world and enjoying a little music while I do it." I laughed back.

"Don't let her fool you. Mimi is one of the smartest people I have ever met, but you all probably know that by now," Clara chimed in with her charming smile and sparkling eyes. She reached across me to grab the bowl from Sherp.

"Now, just how do you plan to save the world, Meems?" Sherp asked as he leaned back into the RV couch.

It was almost dark now. Nik closed the blinds, squeezed in next to Sherp, and laid his hand on his knee. *They are such an unusual pair,* I thought. A red haired, slapdash, unpredictable white guy and a well-dressed, always calm, black guy.

"Well, I want to take down organizations that are knowingly destroying our environment. I want to represent people who are losing everything to the climate crisis against those companies. My parents always said that if you don't like the system, change it. So, I am going to get my law degree and do just that. Go after the bad guys and help the good guys and hope that enough of those actions will add up to ensuring our kids have a planet to live on."

They both looked amused at my response.

Sherp nodded. "Well, okay then."

And that hit me. I really should be studying and not be in this RV right now. It must have registered on my face because Nik looked at me and chimed in.

"Actually, what you need to do is something to let go of some of that stress you are holding. Girl, the look on your face right now . . . You are wound so tight that spring is gonna pop something fierce if you don't take care of it. You have to come climbing with us sometime. We are going to keep asking and one of these days you are going to say yes. Just ask Clara how fun it can be."

"I don't think it's my thing. Plus, I have way too much to do. Also, I am not hardcore like you all," I said.

"Well, Sherp is the only one that is hardcore. He took a year off to get better. Me, I just like to have fun out there, nothing hardcore about my climbing at all. That settles it. You are coming with us next time we go." Nik smiled at Sherp and nodded.

I laughed and offered, "We will see."

Marcel turned from the front seat and chimed in.

"You know, Meems, I wrote a story last year on the impact of music festivals on climate change. They contribute something like almost a thousand kilograms of carbon into the atmosphere and produce nearly a hundred tons of waste per day. I mean, most everyone has to get into a car or plane to get there. And think of all of the plastic cups. I guess at least we are having less of an impact by all piling into this RV," Marcel said while putting a lighter to the end of the glass bowl.

He took a long hit and coughed out, "And, even crazier, when I was researching the story, I heard that there was some talk of festivals being linked to human trafficking. I couldn't find enough information to write about that in any sort of factual way, but it is crazy to think about. We are all there to have a good time, and to think some creeps are buying people there."

"Don't be such a buzzkill, Marcel," Sherp said. "Our vibe is *positive*. Happy topics only from here on out."

I looked around this table. All of us come from different backgrounds. These people are gradually pushing me outside of my comfort zone and my usual routine.

"I have a fun game to play," Clara announced. "Each of us goes around and says what our real-world superpower is and what superpower we would have if we could pick anything."

"Sounds interesting," Kiki said. "I'll go first! Let's see . . . If I could pick any superpower, it would be teleportation. I would love to be able to just zap myself from one place to another. Beam me up, you know! And my real human superpower, well . . . I am like a snake charmer but for people. Instead of a flute, it's something in my voice."

"Ohhh, that was good," said Nik. "My real-world superpower is that I am able to tolerate Logan Sherp and all of his crazy antics." Nik laughed and nudged Sherp's knee. "And if I could pick any power, I would want to see into the future so I could know where to invest my money and make millions."

"Always thinking about how you're going to get rich. You crack me up," said Marcel as he turned around towards the back. "My superpower is that I'm able to read people really well. And if I could pick a fake one, I would become invisible. That way I could go wherever I needed to get the story. Carter, what's yours?" he said and looked across to the driver's seat.

"I think you all know my superpower is my ability to make just about anything grow and most anything taste good, which is a good thing because most of you all cannot cook worth a damn." Carter let out a low chuckle. "And I have to think about what I would want as a real superpower. Maybe the ability to create peace wherever I go. Stop wars and stuff like that."

"Deep. Damn. Deep, Carter." Sherp said. "Well, my superpower is making people laugh. If I had to pick another one, it would absolutely be super strength so I could climb any mountain I wanted to."

All eyes seemed to move to me. I hated being the center of attention. "Well, I guess my superpower is that I am a secret ninja. If I had to pick another one . . . Um, I would want to be able heal mother nature so I could stop climate change." I turned to Clara, the last one to go, but she was distracted by Kiki who was clapping her hands in excitement.

"No way! I am a blackbelt, too. But in karate. I have been training since I was five, a prerequisite of my mother for me to do anything outside of my house. We lived above a dojo growing up, and I helped clean to take classes there. So cool that we have that in common."

"No way! Meems and I here both were forced to go through training to be blackbelts in high school. Our dads insisted that we know how to protect ourselves going off to college," said Clara. "That makes three secret ninjas in this ride. Who would've guessed!"

Clara and Kiki were cracking up laughing, and the rest of us joined in. I wasn't going to let Clara get out of telling us her superpower though, so I threw the small pillow that was on the couch next to me at her.

"Okay, um . . ." Clara started as she fidgeted with her earring. "I guess my superpower is extreme empathy. It's almost like I know exactly what other people are feeling. So much so that sometimes it is not a superpower but more like a super weakness. And if I had to pick one, I would want to speak to animals and be able to understand them."

We rolled on through the night, conversation roaming all over and into Friday morning, finally arriving at the venue just after lunchtime. It was packed with cars as far as you could see. Flags raised over tent and camper compounds to help the wanderers find their way back. A group of men in skirts and girls in bikini tops twirled hula-hoops beside us as we rolled in.

The weekend was a complete blur. Way too little sleep. Some seriously disgusting mud. Hilarious outfits that I think people were legitimately wearing thinking they looked cute. So much incredible music. I barely saw my roommates. It was like swimming in the ocean. The people moved in tides, ebbing and flowing from one stage to the next. I often wandered on my own.

The music had been excellent, but even better was the people-watching. So many doppelgängers! I swore I saw my new political science professor. Thousands of people behind sunglasses and hats. It was impossible to know who was who. I can guarantee that there were famous people in the crowd. I ran into so many musicians in my wandering. They blended in like everyone else, which was sometimes odd. They deserved to have fun like everyone else, but it was cool to see them in person. I was cynical heading into the trip but ended having a great time.

Sunday came and we all piled back into the RV exhausted, ready for the drive home. I finally had the chance to scan my phone. Dad was back from his trip and Mom was headed out to Denver. I let them know I was safe and on my way home. I slept almost the whole way back. I vowed to myself that I was going to lay low for a while, recover, and focus on school.

Back to reality. The final few weeks of school.

I adored my ethics professor, Jessie Holmes. She was such an inspiration to me. Brilliant, articulate, and edgy. A non-conformist. Dreadlocks. The way she was able to take ideas that were hundreds of years old and apply them to today's society was astounding. She was the mentor I didn't know I needed, shining light on pathways of thought that challenged how I viewed the world.

I needed to ace her exam tomorrow, so I was cramming hard. I had just gotten caught up on my laundry. I only have morning classes on Mondays, which is great as it gives me a lot of time to study in the afternoon.

My phone buzzed. I left it in my pocket, trying to focus. And then it buzzed again. *Okay, fine.* I pulled it out and saw a message from a number I didn't recognize. *Meet at The Bean this afternoon?* read the first text. *Coffee on me. 2pm?* read the second. I had no clue who it was so I shoved the phone back into my pocket.

"Hey, Meems. Wow, look at you go, laundry folding champion." Clara laughed as I took handfuls of clothes and stuffed them into my drawers. "I am headed over The Bean to grab a coffee. Do you want one? I need a good distraction from this project I am working on," said Clara.

"Actually, so weird Clara, I just got this random text from someone, must have been a wrong number. They were asking me to meet them at The Bean in an hour." I watched her eyebrows rise and she crossed her arms across her chest. A huge smile broke across her face, a real one, the kind that makes her nose scrunch up and her dimple show.

"What?" I said. "Why are you smiling like that?"

"Well, I was at the Club last Thursday with some peeps from class and Damian came up and tried to mine me for information on you. I told him he would be better off getting the story straight from the source and I gave him your number," she said proudly. "I figured you wouldn't mind. I mean, you are so hot for that guy. It is almost kind of funny to watch. I am surprised it took him this long to text you."

I sat back. *Holy shit.* That was Mr. 10 O'clock, forearm tattoo, hottest guy in this town, texting me to grab coffee. I have zero time to do this. But how can I not go?

"Clara, what do I do? I don't have time for this right now. I have so much to do."

"You don't not have time for this, Meems. You have to go." She smirked. "Oh, and by the way, I might just have my own little romance starting to bud." She looked up at me with her sparkling blue eyes.

"I *have* noticed!" I smiled back at her.

"He is really helping me with my story for this capstone class." Her smile was big and beaming as she gushed. "I want it to be amazing, and hopefully get it published somewhere decent. I haven't heard back on any of the fellowships or internships I have applied for. So, now I need to get serious about finding a job. I started to dig into that stuff Kiki was talking about a bit more. About illegal timber trading. It's pretty interesting," Clara adjusted her large green earring.

Clara had the best jewelry. I barely ever wore jewelry, sometimes the diamond studs my parents gave me for high school graduation. Besides that, I had the same necklace I'd worn for years. A simple silver chain. And a very basic bracelet that never left my wrist. That was about as fancy as I got.

"Don't worry, I am not out chasing bad guys, anymore. At least, not for a while. I am just learning more about how it all works, and it is crazy."

"I am super happy for you about Marcel. He seems like a genuinely good guy. And now here I am going to meet a complete stranger. I can't believe I'm doing this. This is so not who I am. I don't know what this feeling is, but when I see him, it's like I can't function or think of anything else," I said as I grabbed my satchel.

"It's lust. I've been there before. Just be careful, Meems. Have fun and be careful." She winked at me as I turned. I stopped to look at myself in the full-length mirror before leaving. Not much to look at, really. I wear pretty much the same thing every day. Before leaving, I decided to put on a little mascara and some rose essential oil that Clara had given me for Christmas last year.

A BEAN DATE

It was slow at The Bean when I arrived. I grabbed a cappuccino and made my way to my favorite apple table. Then I saw him sitting there. At my table. Of all the tables in this room, he was at *my* table. *What is going on with this world?*

"Hey there, Domino. I was wondering if you would show. You never texted me back," said Damian.

"Well, it is not exactly like I knew who the text was from. You didn't sign it. And actually, I just came here to study for a bit. But since you are here, I guess I can take a break," I said, trying to hide my excitement.

Who was this witty woman talking to this hot guy? I am not funny. I am not quick on my feet. Where are these words coming from?

He brushed his dark hair back from his eyes and asked, "So, tell me about Mimi, the girl whose last name I need to earn to know."

"Nope. Let's start with you. But first, I am so sorry, Damian. I did not piece together that Benji was your brother until you told me your last name the other day. I cannot imagine what you are going through right now."

"It has been tough. I know he had a lot of plans for the Club. I am here trying to figure out what those were. It's really complicated." Damian waywardly glanced down at his cup. "I hadn't ever been here before he died.

I had my own thing going in Miami, working nonstop. The last few years he and I really grew apart. In fact, we hadn't talked since a big falling out about five years ago. I was his only surviving family member, so the Club came to me," Damian tucked a loose strand of hair behind his ear as he took a sip of coffee.

"Your parents aren't around anymore?"

His eyes darkened a bit. He tilted his head to the right and the left and then continued without answering the question, "So, here I am, trying to figure things out. I am not sure how long I will be around; it just depends on what happens with this. And so, that is the gist of me right now. Now, let's hear about you."

He changed the subject fast. I was a little taken aback that he seemed so distant and matter of fact about his brother dying suddenly like that. I guess guys just process things differently. I keep my emotions close to my chest, too, though.

"Not much to share about me. Just a normal girl trying to get through college and figure out what she wants to do with the rest of her life. Pretty boring stuff over here actually," I said.

"I'm not sure I believe that the Ms. Mimi who has not told me her last name, where she is from, what she is studying, or really anything about her, is a boring person. From what I can tell, you have a real knack for critiquing bands, you like the dark corners, maybe aren't much of a crowd person, seem to always be deep in thought, and maybe are a little too quick to judge others. Am I getting close?" He leaned across the table and looked me dead in the eye.

"Well, not really. I spent my whole life being judged by others who did not really know what to make of me. Am I Asian? Am I Black? Am I white? I pretty much pride myself on not being judgmental. But maybe you were right on one thing: the band critiquing. I can't help myself," I said, leaning across the table to meet his gaze.

"Is that right? Well then, let me ask your opinion on this: Do you feel this same electric attraction that I feel between us?" He leaned in closer.

And then, somehow, we were close enough that our lips softly brushed. We stayed there, looking at each other for a second.

I whispered, "It is possible that I feel that, too," and leaned back into my chair.

What in the hell just happened here? Did we just kiss in a public setting? My insides turned inside out. There was a tingling feeling in my stomach and elsewhere. My face flushed. I dipped my chin so my hair fell to cover my cheeks.

"So, I hate to do this, but I really have to go study and pack. Raincheck for another time?" I said, stood, and pushed my chair back, trying to grasp what just happened.

He looked up in a sad, disappointed way. "I guess if that is what you have to do, that is what you have to do. I look forward to seeing you again soon. Before you go, are you going to tell me anything about yourself? How about something small, like where you grew up or what your favorite food is?"

"I guess you can find out my favorite food and get more of my story when you buy me dinner next week."

I gave him a slight smile and walked away. I wondered again where this newfound bravery came from. I have clearly watched far too many rom-coms.

The butterflies continued as I floated down the sidewalks of the neighborhood streets on my way back to the house. My head was so far in the clouds that I almost failed to notice that the magnolia tree blossoms had started to open. They took my breath away. I paused to admire the grandeur of this tree then realized I had been on autopilot. I vowed to not go any further with this random guy; I did not have time for this in my life right now.

I pulled myself out of the trance and walked to the front door of the house. Clara and Carter were reading in the two weathered Adirondack chairs on the front porch. Clara looked up at me and met my gaze. She immediately got the message, stood, and followed me inside. Carter raised an eyebrow and looked at us as we peeled off to head inside. He shrugged then turned back to his book.

I grabbed both of her arms and squeezed them while doing tiny little excited hops. "Clara, I think I just kissed Damian. *In public*. At The Bean."

"Well, well, well, Ms. Mimi Martin finally lets down her guard. This is huge, Meems. Seriously, when was the last time you kissed a guy? Was it Ben our senior year in high school?"

"No, of course not, I have kissed guys since then," I retorted. But in saying it made me realize it really has been since then. "Clara, I do not have time for this nonsense. I am so distracted by what happened that I am going to fail my ethics exam. I have to get an A in that class. I don't have time for guy drama, much less any sort of relationship. Can you imagine what my mom and dad would say if I let my GPA drop right at the end of school? That would be no good at all! We have less than a month until graduation. I just need to stay focused until then."

"Meems, your parents love you. I don't think they would be mad. They only want the best for you. Seriously, you are way too hard on yourself. You study all the time. Give yourself a break," Clara said.

FEASTING

I walked downstairs to a rich aroma of sage and thyme. Carter and Marcel were in the kitchen with a big pot on the stove. My mouth watered. "There will be plenty if you are hungry," Carter said.

"Carter here has been cooking all day. Went out mushroom foraging and has some fresh herbs from the greenhouse out back, a regular José Andrés in our very own kitchen." Marcel smiled and carried a loaf of bread to the oven.

Clara was tucked at the tiny corner bench table hunched over her laptop. Marcel looked over at her and I saw her shyly tuck a strand of her hair behind her ear and smile faintly. They were ridiculously in love with each other. It was so obvious.

I grabbed a glass for water and walked over to the sink. Carter slid around behind me to grab the salt and his chest brushed against my back. I felt an electric shock go through my body and I let out a little sound. It was completely innocent! There was only this narrow walkway between the fridge and the prep island. I turned to look at him and he avoided my gaze. His cheeks flushed red. *Why did I have that reaction?* I must be totally amped because of Damian.

I turned to leave the kitchen, saying over my shoulder, "Wonderful. I would love some. I will be back down in a few." I headed upstairs to grab my laptop, all the while still trying to unpack what just happened at The Bean and then in the kitchen.

I pulled my phone out of my pocket and saw a missed text from Damian. *When can we finish our conversation? Dinner?* He wanted to have dinner tonight. I was just with him a few hours ago. This all seemed really fast. For me, at least. But a primal urge was taking over my body; there was no answer except yes right now. I texted him back, *How about dessert, 8pm, the Chocolate House?*

Downstairs, I saw Marcel slide in next to Clara on the small bench with two bowls of soup. The table was set up in the corner with small benches built into the wall on two sides and two small chairs in the other corner. A beautiful picture window looks out into the yard across from the table and two small windows sit above the table, letting in the colors of a beautiful sunset.

I saw that Carter was holding two bowls. "I made you one," he said, smiling at me.

He was such a strange bird. He's a brilliant surfer. He is also thoughtful and deeply philosophical, and apparently an incredible cook. If he would only lose that cliché hemp necklace. And the socks. I cannot deal with the socks and sandals. Maybe a haircut. Then he might have something going for himself.

"Yes, thanks." I grabbed one of the bowls. Our fingers brushed and that same electric shock ran through my body. My eyes raised up to meet his. Time stopped. It was like he was looking all the way into my soul. I shook my head and stepped back and moved over to the table. I was totally imagining that moment, right? I am into Damian right now. That must be bubbling over into other parts of my life.

As always, Clara and Marcel had us laughing the entire meal. They were hilarious storytellers. They were talking about going to "scope out some joints" for Clara's story tomorrow. They make a good pair.

I felt my phone buzz and looked down. I had completely lost track of time. I was supposed to be at the Chocolate Factory in 30 minutes. I thanked Carter for the meal, gave Clara a little wink, and put the dishes into the dishwasher on my way out.

RED, RED WINE

The Chocolate Factory is a dark little spot with small metal tables and wooden mismatched chairs. Candles provided dim lighting on each of them. A case with freshly made desserts stood in the front of the room. Damian was already there when I arrived. There was a bottle of wine on the table. I smiled and walked over as he poured me a glass.

He stood as I came closer to the table and extended a hand towards me. It felt so natural. I reached out and took it and noticed how soft his hands were. He pulled me in for a hug and the internal fire was lit, burning so hot I could hardly contain myself.

The conversation was easy and flowed well. There were no pretenses or pressure. It was like we had known each other forever; except I had never known anyone with a voice like his. Silky with a slight accent from somewhere in Latin America, I guessed. It was very subtle.

We talked about the state of the world and climate change and science fiction. I leaned in to ask him about where he grew up and my knee brushed his under the table. I felt his hand on my thigh. I couldn't contain myself any longer. I put my hand on his and stroked his fingers. He looked at me in a way that no one ever has.

He grew up in many places, depending on where his father was able to find work. I learned that he was originally from Brazil and lived there until his teen years and then moved to Mexico. He had been in Miami for most of his adult life.

He spoke so softly and gently. His brown hair fell across his forehead in gentle curls. He leaned forward and brushed my hair back behind my ear, eliciting electric jolts through my body. I tried to remain calm and steady my breathing.

"Tell me about this tattoo. It is . . ." I looked at his forearm that was resting across my lap and then at him, "very curious."

"Why don't you tell me about you, mysterious Mimi?"

We had barely touched our desserts when the server came over to tell us they were closing. The server handed us the bill and asked if we would like our desserts boxed up. We both shook our heads no. Damian placed his credit card on the check and the server walked away.

"Would you like to continue our conversation back at my place?" Damian asked.

Everything ran through my mind, but the only thing clear answer was yes even though I knew I should say no. Going back to his place could only lead to one thing.

"I have an early day tomorrow. I'm not sure it's a good night for that," I managed to force out.

"No pressure. I was just really enjoying myself and getting to know you. We can do it another time," he said and grabbed my hand as we stood up to leave.

I turned to him as we walked out to the parking lot. He was so close. It was magnetic. His body was warm against mine. His lips soft, his caress behind my head was gentle. I let out a low moan as I leaned into his arms.

"Maybe I can come over for a few minutes," I whispered in his ear.

He opened the car door and I climbed in.

"I can bring you home whenever you need," he smiled at me, and we drove the few miles outside of town to his apartment. He lived on the top floor of a beautiful old house not too far from where I lived.

As soon as we entered the door, I lunged for him. My whole body longed for him. I had never had this feeling before. It was the lust that Clara had talked about. It was so out of character for me. We fell into a deep embrace. He pulled slightly back, smiled, brushed the hair behind both of my ears and held my chin in his hand. "I certainly had not expected to find someone like you in this small town." He stepped away to the kitchen. "Let me get us some wine. Make yourself comfortable."

I took a deep breath, straightened my clothes, and walked around the living room, noting its sparse decorations. The molding, fireplace, and woodwork were the décor. There were a few books on the shelves, one simple couch, and no television. A desk with a laptop on it sat in the corner. Damian walked back through the wide high archway separating the living room from the dining room. He handed me the glass of wine and we sat on the couch.

"So, Mimi, are you going to tell me anything or do I have to keep guessing? How about three things? Any three things I don't already know," he said.

"Okay. Things you already know. I am trying to get

into law school. I love live music. My best friend is Clara. Things you don't know: My mother is a famous novelist. I have a second-degree blackbelt in Tae Kwon Do. I have traveled to over 20 countries." I set down my wine glass. "Actually, I think I need to go Damian."

"Okay." He nodded and also set down his glass of wine.

And then I leaned in and kissed him. We fall back onto the couch, our bodies perfectly fitting together like pieces of a puzzle. He caressed the skin along the side of my rib cage down to my pant line. Then he pulled off his shirt. He was incredibly fit, surprisingly so.

He untied my shoes as I laid back on the couch then came and sat up over me. *How did this happen? How am I here with this god-like man? I know he is at least five years older than me.* This was crazy but I couldn't stop. While he untied my other shoe, I found his belt buckle and undid it.

He looked straight at me, saying the unsaid. He slid my shoe off and I rose up to meet him, wrapping my legs around his waist as he carried me back to his bedroom.

My back touched his sheets. They were cold. I struggled to feel anything besides desire. I felt transported to another world to pleasures I didn't even know existed. It felt like we spun in circles for what seemed like hours. I was exhausted.

Afterward, lying together on his platform bed, he gently stroked my arm. I didn't want to leave, but knew I had to. I couldn't stay here. It had to be close to midnight at this point.

I have only ever been intimate with one other person. We'd dated for over a year before we had sex. I was 16. It has been five years since then. I am not the kind of girl who sleeps with random strangers. But this felt genetic, the attraction I have to Damian, like I was programmed to want him, and I couldn't control it. Which is exactly why I needed to leave . . .

I rolled over and pulled the sheet around me. The self-consciousness I lacked earlier came back with a vengeance. I didn't want him to see me naked. I looked around for my clothes. They were scattered everywhere.

He propped himself up on his elbow. "Do you have to go?"

"I'm sorry, Damian. This is not me. I am not the type of girl who does this kind of thing. I need to go." I grabbed my bra from the corner of the bedroom. "I have a test tomorrow and I just need some time to think. Don't worry. I'll call an Uber," I said as I slid my underwear towards the bed with my foot.

"Nonsense, I will drive you home. Here, let me help you," he said as he slid out of the bed. I glanced over my shoulder at his back where I admired his sculpted, pronounced shoulder muscles. He was thin, lean, no fat on him whatsoever. There was a scar that ran from his right scapula down towards his chest.

Damian collected my t-shirt and jeans and brought them over. He stood in front of me, confidently nude. He bent down and picked up his jeans, all the while looking at me with a slight smile.

"I will give you some privacy while I find my keys."

He walked out of the room, and I could finally breathe.

I felt like I was sneaking into my parent's house. I slowly opened the door and hoped it wouldn't wake my roommates. It was a weeknight night, after all, and I knew most everyone had early classes tomorrow.

I slid through the door and turned to slowly close it, only to find Sherp, Nik, and Marcel sitting on the couch watching a movie. They all turned and looked right at me. Nik gave me the double eyebrow raise two times in a row, signaling that he knew what I had been up to.

"Hi, sorry, guys. Late night. Um, gotta go to bed. Bye." I tiptoed up the stairs and thanked the universe that Carter and Clara didn't hear me.

I set my alarm and crashed into bed, exhausted from the erotic night. But the energy in my body made it nearly impossible to fall asleep. Next thing I knew, it was 7 a.m. and my alarm was blaring. Reluctantly, I dragged myself out of bed and got ready.

Clara and Marcel were already in the kitchen having coffee by the time I came down. They were talking about their final journalism projects, which interviews they needed to conduct, and what research was needed.

I knew I also needed to buckle down and study for finals, which were only two weeks away. Today, my day was going to be spent at either the library or in class. No distractions.

But Clara got to me. "So, got home late last night,

huh?"

"Nope, not too late, gotta run," I said and grabbed my mug from the counter, filling it with her sludgy coffee.

"Uh-huh, we can talk later," I heard her say, giggling as I tried to get out of the kitchen as fast as possible. There was no way I wanted to have that conversation in front of Marcel.

I got to the library around mid-morning and turned my phone off. I didn't want Damian to text and distract me even more. Would he even call? Was I just a one-night stand? I really needed to get all my exam prep in order, make flash cards, outline my notes.

Fortunately, I was able to get in my zone and focus hard on my studies. Around 5 p.m., I realized I was starving. I'd unintentionally skipped lunch and had to find somewhere nearby to get some food. The student center had terrible food, but it was close, and I could use the meal card.

Shoving books into the bag, I decide to run over there and come straight back to the library. I decide to turn my phone on so I could listen to some music on my way to the student center. A little bit of a reset and brain break. Big mistake.

I had two missed calls and a text from Damian.

The text read: *I had a really wonderful time last night. I hope we can get together again soon. I will be at the Club if you want to stop by.*

And the voicemail I could have listened to 100 times. "Mimi, thank you. I haven't felt that good in a long time. I hope I can see you again soon. Call me."

All of this had me utterly distracted. I was never going to be ready for finals week. I decided to go home and try to finish studying there. If nothing else, get a good night's sleep. It didn't take long for the long day to hit me, and I fell asleep reading my philosophy homework.

I woke up with a start. My class! I found myself spinning into an existential crisis as I jumped in the shower and hurried to get ready for class. This morning was philosophy class, a great place to sort out the details of humanity and what it means to be human. Just what I need.

I found my way to my normal spot, midway up on the right aisle. Not too close, but still close enough that the teachers know I'm paying attention. And I know they can see me, so I have to pay attention.

Today, Dr. Holmes talked about David Hume and if we can really ever achieve a true conception of ourselves. I half listened as my mind wrestled with this idea.

After class, I approached Dr. Holmes as she packed up her bag. "Dr. Holmes, do you have a minute?"

"Of course. I actually have 15 of them before I need to head to my office hours. You are also welcome to join me there. What can I help you with, Mimi?"

"Well, I think I am having a bit of an existential crisis. Struggling a bit. I keep thinking about how I thought I knew exactly who I was and what I was going to do with my life, only to realize maybe I am becoming someone else, and that someone is not sure what trajectory she should be on." I paused for a moment, realizing I blurted all of this out without hesitation. "Are there any good books you can recommend? Surely, the great thinkers have processed a good existential crisis in writing." I couldn't believe what I just shared with her. I hadn't even fully thought that through to myself, it just kind of fell out of my mouth.

"That, my dear, sounds like something a good therapist can help you work through. And if you really want to wrestle with the philosophical demons on this one, there are a few." She smiled at me patiently. "If you want to think about the very essence of who we are, for that you can read my main man, Aristotle. Or maybe you feel a moral duty or moral obligation. For that, Kant is always on your side. Or maybe you are more of a Jean-Paul Sartre gal, in search of experiences to be authentic in life.

"Point being, Mimi, we are all on a constant journey. That is part of being human. We all struggle. It is why all of the greats had something to write about. If we can learn from those struggles, we are better for it. Maybe it is not your time to solve this. Maybe you need to struggle through it with an open mind and see where the journey leads you." She smiled at me.

"Thanks, Dr. Holmes, this is just what I needed to hear." I nodded at her, smiled, and turned to walk up the stairs towards the door at the back of the classroom.

"And Mimi, you can do more than one thing in life. You can and should be absolutely more than that one thing you do. It does not have to be who you were and who you are now. It is *all* you. You are just becoming more you with time. And you define your own path. No one else should do that for you."

As I left the classroom, I felt the weight that had been sitting in the pit of my stomach disappear.

I walked over to The Bean to grab a coffee between classes and decided to call my mom.

She picked it up on the second ring. There was a lot of background noise that sounded like she was in the car. We exchanged a few words and I let her know I might come home for the weekend.

I felt butterflies fill my stomach as I entered The Bean. I noticed that the place was mostly empty. No one was at my apple table. No sign of Damian. I feel a sense of relief.

It's Thursday now, three days since my magical date with Damian and I haven't talked to him since. He has texted a few times, but I honestly have no idea what to say to him and my nose is buried in books right now, so it has been easy to ignore him.

Walking back from campus after my last class, I decided on a whim to stop by the Club. The doors were locked, which was not surprising as it was only noon and they did not open until four. Damian must have heard the door rattle, though, because he came around the corner and looked very surprised to see me. He came forward and unlocked the doors.

"I was beginning to wonder if I scared you away," he said gently, looking deep into my eyes as he reached for my hand. He led me inside and locked the door behind us.

"I don't get scared, Damian." I smiled. "I have just been busy with school. I wanted to stop by because I am going out of town this weekend. I'm hoping we can get together when I get back."

We made some small talk about what bands were playing soon and how business had been. It was easy to talk to him, but I wanted to do more. My knee brushed against his as we sat on the barstools at the edge of the long bar. He pulled me in close and kissed the side of my neck. I leaned in and he stood up, standing between my legs. I wrapped my legs around his waist as he continued to kiss up and down my neck. He slowly lifted me up onto the table behind us. In an instant, it was like the cosmos had clashed and a new universe was created. It was carnal.

I was glad that I wore the lace bra today. He stripped my jacket off and pushed my shirt up over my head and then proceeded to take his off. His hips moved in a way that only Latin hips could move. I suddenly came to my senses and pushed him away. I couldn't do this here, now. He stopped and pulled back.

"I'm sorry," he said. We took a moment to catch our breath. He put his hand on my leg as I sat up on the table. "When I saw you, I was just completely overwhelmed. I have been thinking about you nonstop for the last few days."

My hair fell across my face as I smiled. "Yeah, I understand. I feel the same way. But I need to take this slow. I like you. A lot. I mean, we clearly have chemistry, but we barely know each other. How about I call you when I get back in town and we can grab coffee? Or I will come by here to see some music one night." I pulled my shirt back over my head and slid down off the table. I honestly didn't know that I had this willpower in me.

"Of course, I would love to see you whenever you have time. And, just for the record, you know me better than anyone else in this town." He smiled and handed me my jacket, kissing me on the cheek as I turned to leave.

I didn't have class tomorrow, so I decided to leave tonight to see my parents for the weekend. I love that my parents live close. But not too close. About an hour away from Lexington, on the east side of Louisville. Pretty much a straight shot down I-64. I headed back home to get packed.

THE OLD HOMESTEAD

Niles had invited David to come by on Friday morning to reconnect and get organized before their upcoming trip. He and David had a long history together, but it had been years since they had seen each other. They recently reconnected through working on this human and timber trafficking case.

The soft morning light streamed through the translucent curtains. The light pink hue indicated that the sun was just beginning to rise. Hae was in bed quietly snoring. He had no idea what time she had come to bed.

He grabbed his flannel robe and walked downstairs. One thing he would never give up was the printed newspaper. He loved the tradition of opening it up with his cup of coffee. It was not his only source of news, but it was a nostalgic and cathartic routine. As he opened the door, he realized Mimi's shoes were sitting next to it. A welcome surprise. She must have come home last night after he fell asleep. He must have been really out to have not heard her. His heart warmed at the thought of her being home, and then he remembered she was home a day early and that David was going to be here any minute.

He stepped outside to grab the paper and saw the black SUV parked down the road. Surely David had seen him come out and would soon come to the back door.

Their kitchen was old enough to still have large wooden swinging doors that separated it from their dining room. Niles went into the kitchen, popped the wooden door closed behind him, unlocked the back door, and turned on the coffee machine. Just then, David appeared at the back door. Niles motioned for him to come in. It had been a long time since they'd seen each other. The handshake turned into a bear hug with the two slaps on the back.

"Coffee will be ready in about five minutes," Niles said. "I can't talk business until then, so you're going to have to catch me up on your life from these past few years."

"Still with DHS, as you know. I need you to know that Mimi knows who I am. She did not remember me from those many years ago when she was a child."

"When did you meet Mimi?" Niles asked, caution rising in his voice.

"I stepped in temporarily as a political science professor at the university. She is taking one of my classes. I must say, she is a very bright girl, and a fierce debater. She really impressed me. I approached her after class to see if she recalled who I was, but nothing," David said.

"Is this academic adventure your next big step in life or are you staging for the case?"

"Being at the university allows me all kinds of connections and access to events and social circles. I'm just a visiting professional who stepped in temporarily to help out a colleague. It is not so different than our days back at Georgetown, honestly."

The coffee pot chirped its three-bell ring. Niles poured them two cups, remembering that David and he both drank it straight.

"Listen, David, something is off. I know you've seen the reports. The woman I met with at the land tenure NGO in Vietnam was shot just after I left. Clara was shot at while scoping out a sawmill. While the two events seem completely unrelated, I can't help but relate them in my head."

"Have you heard any more details on what happened in Vietnam?" David said quietly.

"I have no details. I just got the call from the Agricultural Attaché as I was leaving the country. He was checking in on me," Niles said in a hushed tone. "I am also pretty sure I was being followed most of the time I was in the country."

"That is actually not that odd. Vietnam is known for keeping an eye on their guests."

Niles and David discussed the details of what they knew about Brazil so far. It was the next stop for the case and this time they would finally get to work together again, in person. The leads looked promising.

"There is more, Niles. Yesterday, just outside of Lexington at one of the rest stops, police raided a truck carrying almost 20 women and girls. They were primarily from Honduras and Brazil. We are still investigating. This is a lot of activity for sleepy little Kentucky." David sat back in his chair.

"What do you think about the mill where Marcel was shot? Any leads there? Any connection to the case?"

"Not yet, but Anne is looking into it. She is one of the best special agents in charge that I have ever worked with. If there is something to find, she will find it," David said.

Right then, the wooden door swung open, and a half-asleep Mimi walked through.

"I smelled the coffee and couldn't resist," she said as she headed to the cabinet to grab a mug. It had not registered to her yet that someone was sitting in the kitchen with her father.

Mimi slowly turned and then her eyes widened. "Professor Pile. What-um . . . Hi. Dad, why is my political science professor in our kitchen?"

"Good morning, Mimi," David said as he took a long slow sip of his coffee and turned his gaze to look directly at Niles.

"Mimi, I can explain," Niles said. "David is an old friend of mine. I just learned he was in town and he stopped by for a visit. He mentioned that he had put two and two together, that you were in his class and that you were a star student." He flashed a smile as he took a sip of his coffee.

Niles set his mug down and went to greet his daughter. "Now, doesn't your old man get a hug when you come to visit?" He embraced Mimi. But she was still in shock that her professor was sitting in her kitchen at six in the morning on a Saturday.

"Um . . . Okay, I am clearly not awake enough for this yet. Professor, good to see you. Dad, I am going back upstairs and will let you two catch up." She took coffee to go and retreated back to her room.

Niles turned back to David. “We need to be careful here. I don’t want my family involved in this. This was supposed to be a straightforward case. Do a little background research, talk to a few people, figure out some sources. If things start to get complicated, I need to get out,” Niles said, looking David straight in the eye.

"Niles, we have a team of agents here, and an entire department in DC working on this. We will look out for you and your family. You are one of the best intelligence officers I have ever met. You see things other people can’t. You find information everyone else misses. People talk to you when they won’t talk to anyone else. We need your help on this one.” David rose to leave.

“One more thing. Mimi’s roommate, Kiki. I have looked into her research. She is really breaking new ground in this field, and I think could be a tremendous asset to the case. I wanted to let you know I may contact her to help trace some of this wood we just seized.” David patted Niles’ arm as he moved towards the back door. “It was really good to see you, old friend. I will see you in Brazil in a few days. But next time over a beer, not at sunrise, okay?” He nodded at Niles as he exited out the white wooden backdoor.

Niles watched him go. He noticed that a soft fog lifted off the grass as he walked. Niles sipped his coffee and thought about what David had told him. Things were happening here. Very close to home. Too close to home.

Niles finished his coffee and walked upstairs to shower. He noticed Mimi’s door was cracked open and decided to pop his head in. She sat cross-legged on her bed, looking at her laptop with coffee in hand. He'd seen her like this a hundred times over the years, except without the coffee. It felt like home, having her here.

"Hey, kiddo. Sorry if that caught you off guard. David was passing through and knows I'm an early bird, so he stopped by," Niles said.

"Not weird at all to come downstairs at six in the morning to find a complete stranger in your kitchen. But wait, he is not a complete stranger, he is my new professor. Nope, not weird at all," Mimi said kind of laughing, kind of not.

"Well, he hit the road a few minutes ago. So, I'm free. Want to go get breakfast?"

"Actually, let's just make breakfast here, Dad. I kind of just want to hang out at home today, if that is okay."

"Sounds wonderful. I'm going to grab a shower. Then I will make your favorite, French toast." Niles walked out the door and started his routine as if it were a morning like every other.

DAMIAN MEETS HIS MATCH

The books were just not adding up. Damian ran the numbers again and again. His brother was a terrible accountant, that much was obvious. He paid way too much for the remodeling of this place. Damian was no financial whiz but knew that he needed to either start fresh or dig in more. He couldn't sell this place with financials like these. He'd been putting the word out on the street that he might be sell to the right buyer, so he was trying to get everything in order.

It was a Sunday afternoon, and the Club would be open in about 30 minutes. His main bartender, Tim, had not yet arrived. The rest of the staff usually came in between 5-8 p.m. Business didn't really pick up until after 8 p.m., depending on the band and the night. They didn't have a band tonight, so it would just be him and Tim working. They probably wouldn't have more than 20 people come in all night. He started to think he would close on Wednesdays and Sundays, too, to save some money.

He heard footsteps behind him as he stocked the bar. He assumed it was Tim. He turned to say something and realized three guys, not part of his normal crowd, stood in front of him.

"Hey, guys, sorry, not open yet. Give me about 30 more minutes to get things in order, and then I will be happy to help you out. Until then if you don't mind,"

Damian nodded towards the door.

"We know you're not open. We're not here for drinks," the short one, who was quite robust in the stomach and had what sounded like a Russian accent, said. "We are here to talk business."

Damian stayed behind the bar. The gun he kept for emergencies was all on the way on the other end. There was no way he could casually make his way to it.

"And what business might that be?" Damian asked, trying to sound casual.

"You lack understanding of how this works. We were giving you some, um, let's call it 'space,' to get settled. But a deal is a deal. Benji made us a deal, and you will make good on that deal." The taller one, who had tattoos all over his body, stepped forward.

Damian put his hands in the air. "Whoa there, fellas. I'm not sure what deal Benji made you, but I can tell you that I want no part in it. I'm just here trying to settle my brothers' affairs and move on with my life. I have no plans to keep this Club."

"This is not a choice that is yours to make," the short fat one said, his hand on his hip. "The Club is part of the business, and now you are part of the business. You work for us. We will be back in a few days to go over everything." The three guys turned and left.

Damian stood there dumbfounded. He had no idea what just happened. What had his brother been into? They must have had a key to the door. Damian was sure the door was locked. They came in here like they owned this place. Damian quickly realized he made a big mistake by coming to Kentucky.

He turned around and looked at the video cameras in the top corner of the room. It reminded him he also had exterior cameras. He was going to look and see if he could get car plates and see who these people were. He was sure they would be back, and he wanted to be prepared.

He went to the back room. Sure enough, there it was. Not the brightest criminals, apparently, parking right out front where the cameras and everyone under the sun could see them. He grabbed the plate and car make and model and did some research. At least he knew what to keep an eye out for. He knew better than to go to the police.

He realized in looking at his phone that he had not called Mimi back. His whole being here was way more complicated than he'd imagined. He thought he'd come to Kentucky, figure out the value potential for the bar, and get back to his life in Miami. Now, he was apparently in deep with something he knew nothing about.

SUNDAY DINNER

The weekend was over way too quickly. Mom had some book club thing on Sunday afternoon, so I left about the same time and texted Damian as I hit the road, hoping he would respond by the time I got to Lexington.

I have a break tomorrow from 9-945, want to meet me at The Bean?

The drive went quickly. It was a beautiful spring day. All shades of green you could imagine popped from the trees as they greeted the spring sun. I pulled up to the house and was super lucky to find a parking spot right out front. I had texted Clara to let her know I would be home in time for family dinner night. Carter was cooking, so it was going to be really good.

I opened the door to an incredible aroma. Indian food. Nik and Sherp were on the couch playing some video game. Kiki was deep in a book. No one seemed to notice when I entered.

Clara came bounding from the kitchen. "Carter taught me how to make bread. Me, Ms. Domestication, right? Can you believe it?" She kissed me on each cheek and turned on her heel back through the dining room and into the kitchen.

I walked up the stairs to my room, dropped my bags, and flopped down on my bed. It felt good to be here. I thought I needed a break from this place, but it was feeling increasingly like home. I scrolled through my phone. Damian had not responded.

I heard a chime. It was some sort of Tibetan bell, two metal semi-circles that clang together and make a peaceful, yet undeniably loud sound. It reminded me of how yoga teachers brought their students out of savasana. It rang three times. So much better than yelling that dinner was ready. A nice Carter touch.

I dashed downstairs and helped bring the food out to the table. We gathered like one big, weird, diverse family. Only a few more weeks and half of us were graduating. This normal may not last much longer. It had only been a few months, but it seemed like we had all lived together for years.

"First topic of the night: best book you read this year. And go," Kiki said.

Sherp laughed. "The only books I read are climbing books. If you like a good climbing adventure book, then you should read one called something like *Epic.* It's about climbing high peaks."

This conversation continued almost all the way through dinner. These were some prolific readers.

Dinner finished up and I offered to help with the dishes. Carter followed me into the kitchen. "Did you have a good visit with your parents?" he asked.

"Yeah, it was really chill. I studied a lot. We made some good meals together. Went for a walk."

"What I wouldn't give to be able to do that kind of stuff with my parents again." His voice sounded sad. "Enjoy it while you can, you never know when things will change. Out of the blue, something happens, and it's all different."

I realized that I'd only spoken about his parents in passing before. He stood there holding plates, looking out the window. I walked over to him, took the plates out of his hand, and set them on the island.

"You okay?" I asked.

"Yeah, sometimes it just hits me. I think I am fine, and then, well . . . Sorry."

"There is nothing to apologize for. Do you want to tell me about them?"

"No, not right now. Maybe another time," he said and reached for the dishes.

"I got this, Carter. You cooked, so you don't have to clean." His hand was on top of mine. He looked at me with sad eyes, his long sandy blonde hair falling into his face. His eyes were green. How had I not noticed that before? They were quite striking, actually.

He pulled his hand back and leaned against the counter.

"You grew up in Louisville, right?" he asked.

"Sure did. Well, mostly. I was in DC for a bit when I was younger, but I call Louisville home. Clara and I lived down the street from one another, best friends since elementary school. Could not wait to get out of that town. Funny how going an hour away can feel so far sometimes, and so close other times." I rinsed the dishes and loaded them into the dishwasher

"I would have never thought to live in Kentucky. My parents always loved it here, the rural feel. I am more of a California guy. But, hey, beggars can't be choosers. I will probably head back out west once I finish my PhD. At least, I hope I can find a job out that way," he said.

"I haven't spent much time out west. Just a few long weekends to visit National Parks and that is it. I will have to come visit you if you end up out there."

"I would really like that," he said. I felt something stir inside me. It was a different feeling than with Damian.

"Okay, off to the grind. This thesis isn't going to write itself." He turned and walked back through the dining room. I turned to look and spotted Clara leaning in the doorway. She gave Carter a look and then gave me her signature one eyebrow raise to say, "What the heck, Meems? Tell me everything."

I just shook my head at her but found myself smiling from the inside out.

TREE FORENSICS 101

Clara, Marcel, and Kiki sat at the large wooden dining room table discussing the ins and outs of tree forensics. Clara decided to create a thread in her story featuring Kiki as a woman who, against all odds, is changing the world through STEM. Marcel just listened. The way Clara saw the world was mesmerizing to him. She had this perspective that brought out the best in people. She was the light in the dark. She could bring a smile to anyone, anytime. And Kiki was absolutely loving the attention, and loving telling her story, she was clearly passionate about this work.

"It is a little-known fact that each tree is completely unique and has a distinct chemical signal. This unique signal allows forensics people like me to trace them to their point of geographic origin," said Kiki.

Kiki came to college on a full scholarship after winning a high school science fair contest by designing genetic assay tests that could identify which flowers had been visited by bees. Her good grades and winning project were enough to get her almost a full ride through her undergraduate education. Her master's degree turned into a PhD, also fully sponsored. She collaborated with scientists all over the world on the technique, building on existing research out of federal laboratories and several private institutions.

She told them about how she had forged a

collaboration with scientists from China, Brazil, Germany, and the Netherlands over the last two years and that she was leading the project. She was going to give her first international presentation soon.

"What is even cooler is that we thought about how NASA does that cool project where any citizen can look at star maps to help to find a new galaxy," she said. "Well, we took that idea and made a database that is open source. Anyone can contribute. People from all over the world are loading their tree DNA and samples into it. The more they add, the more likely we are to be able to look at a sample and indicate the exact *Genus, species* and where it came from."

"So, how do you use the information? I mean it would be really cool to know where this dining room table wood came from," Marcel asked, tapping his fingers to the deep walnut-colored table.

Kiki responded, "It is used for certification-type stuff, like knowing if wood really came from where people say it comes from. And, honestly, it might be used even more as a basis for other science, like people doing ecological research. They could tie in their data with ours to tell a bigger story."

"It was so cool being in the lab with you last week and getting those awesome photos of you all science'd up. It was perfect." Clara beamed at Kiki.

"Well, friends, I need to get my stuff together. I will arrive in Brazil in under 48 hours and need to finish writing my talk. I don't know if all of you know, but that grant I applied for from the Girls in STEM non-profit funded me. So now I get to go on a three-day excursion before the conference to learn about forest practices in the Amazon. I am so freaking excited!" Kiki got up to go to her room.

"One day, when I'm journalist, I will get to travel to exciting places like Brazil. For now, I'm stuck here in Lexington. At least, for a few more weeks." Marcel looked at Clara and smiled. "But I guess it's not a terrible place to be stuck."

Clara looked at Kiki as she was walking away. "I know you will do amazing in your talk. Just do it like you did here, let your heart do the talking and you will be just fine."

BRAZIL

David called Niles to let him know the team on the ground in Brazil found some promising leads. They needed to get down there soon to talk to possible informants. The agents and local partners had done great work stopping some of the *cortador noturnos,* the "night-cutting poachers." However, recent government policies supporting more timber harvesting and clearing of land seemed to embolden the activities.

Loggers are like gypsies; they move from camp to camp to get work. You can almost trace the path of logging camps because the loggers tend to leave sickness wherever they go. Several of these remote areas have never experienced these kinds of diseases. Not only are these loggers destroying precious ecosystems, but they're also decimating local populations when they pass through. Some are even known to also leave a path of moral destruction, taking their pleasures with local women and girls. Sometimes, the loggers take the women and sell them at the ports to the highest bidder. Human trafficking is increasing at Brazilian ports despite authorities working hard to crack down on the entire supply chain.

Niles is aware of all of this. He knows that legitimate logging operations exist and are an important economic engine, but these illegal activities are harming everything. It seems like there is a corrupt aggregator operating in the area, paying top dollar for the rare woods, stimulating more illegal operations.

Niles and David will travel to the harvested areas, talk to the villagers, and also visit legitimate operations, who are losing public trust and support because of the negative press.

David and Niles grabbed separate flights and arrived a few days apart of each other. David arrived earlier to meet the advance team and talk to agents in the area. Niles continued to work his angle as a non-profit consultant.

The plane ride to Manaus International Airport was pretty brutal. Louisville to Houston to Panama City to Manaus in about 15 hours with no time to sleep on the plane. And the turbulence between Panama City and Manaus was stomach wrenching. Niles was glad to have landed. At least the weather was somewhat pleasant. The rainy season just finished and everything was alive. And very humid. Having spent enough summers in the Ohio River Valley, Niles thought he knew humidity well, but the humidity here was a special kind of sticky.

As he came out of the airport, he saw his driver was waiting for him with a handwritten "MARTIN" sign. The driver wore a clean pressed t-shirt and jeans with a Panama hat and a big smile. The driver, Ricardo as he introduced himself, greeted Niles and welcomed him to the city. They got in the car and headed to the Tropical Executive Hotel near the Ponta Negra Beach, situated on the banks of the Rio Negra. Niles admired the soaring hotel with its any windows and lush greenery.

Niles looked forward to getting settled in and washing off the day's travel. He was supposed to meet David and their partner from the Brazilian Institute of the Environment and Natural Resources (IBAMA) for dinner. It was going to be a busy week.

Dinner at the Tropical was typical of this area. Pao de queijo was abundant and dangerously good. These little Brazilian cheese balls are famous. Usually they are served for breakfast, but the tourists love them and so most of the hotels serve them with every meal. The three men ordered their food and then dove into their conversation.

Paulo Santos worked for IBAMA since its inception in the late 1980s. With 30 years of investigations under his belt, he had seen just about everything, including the tremendous and devastating growth of illegal logging in the area. All three men sipped Brahmas, the only beer on draught.

As the food arrived, Paulo described the changes he'd seen over the last few decades. "Lately, land clearing for agriculture is most alarming. The value and demand for beef and leather is driving heavy deforestation. Most of the land is being cleared illegally and without permits. Big, big losses of forests. Very sad for local people, and very sad for the forest." He finished the last gulp of his beer.

"Tell me more about where we are going this week and what to expect out there," Niles said and took another bite of his fish.

"Up to the north there is much activity in Roraima towards Suriname. Here, it is the Indigenous who suffer the most, but they fight hard. I fear for their future. Much longer and there may be none. But most is to the south, right now. That is where we will focus, here there are many people who know things and can help us." Paolo flagged the server for another beer.

"One of the villages we will pass through might be hard. Many men went with the company to cut the forest for money but did not return. There is much anger. They thought it would be a few trees, but the spirit of the forest was cut, and nothing left and their lives badly impacted and husbands gone. They are now slaves to the company," Paulo said.

David chimed in, "It will be a few days' travel inland from here. We will stop at a few places along the way, bring lots of bug spray and mud boots."

They ended dinner with cachaca and brigadeiro. They finished discussing the details of the trip.

The next morning came early. They woke before the sun and met their driver, the same from the airport. Ricardo was waiting for them patiently at the front of the hotel. Niles, David, and Paulo hopped in, and they drove off towards the docks for their trip up the Amazon River.

Until recently, most logging activities took place on the eastern side of Beruri, a small village about 130 nautical miles away. Given the increased access and travel, the Trans-Purus region, which stretches towards Peru, has seen a boom of illegal activity. Ricardo shared that he was part of the ground team. He was an ecologist by training and told Niles and David that this region was one of the last frontiers, as well as one of the last intact havens for biodiversity.

Paulo added, "The biodiversity of the whole Amazon depends on us being able to maintain this block. Many species in the area are threatened, and more than half of the tree species are soon to be on the endangered species register. But it is really the climate. We need to care because of the climate. These forests store so much carbon. To lose them would be bad for the whole world."

"We think we know who has been building the access roads into the area. We will ask the locals," Ricardo said over his shoulder as he drove up to the dock.

All of them wore bulletproof vests. Ricardo and Paulo carried guns. IBAMA agents are not liked in certain areas but celebrated in others. The Indigenous and small communities suffered when developers, gold miners, loggers, and others colonized the land. Nowadays, they'll do nearly anything to defend their property against anyone who seems a threat.

"Small-scale agricultural clearing and illegal harvesting were impactful, but what we are looking at here is a well-funded and organized group. We are seeing the same equipment being used in multiple places, and hearing the same talk in villages," said Paolo as they walked from the jeep to the large boat.

About seven hours into the boat ride, they had traveled the Rio Negro to the Solimoes River and were headed towards the split to the Purus River just outside of Anama. A sleek ferry boat passed the group on the main river headed to Coari. A pod of river dolphins surfaced just before Anama. Finally, they arrived at Beruri, a town of about 20,000 people, and grabbed some dinner. They also met up with a member of the local team with another small boat there.

"There are over a dozen small villages along this river corridor, all would have their own story," Ricardo shared as everyone stepped onto the wooden boat that would be their lodging for the night. The boat had two levels and was layered with hammocks, their lodging for the night. "We will leave at first light."

Around a small table, they reviewed their plan for tomorrow. "The first town, Surara, we will not stop. IBAMA raids last year. A bigger town. The mill closed and people out of work, very unhappy," said Paulo. "But another two hours, there," as Paulo points at a map indicating they would continue to travel downstream, "the men left as slaves, and the women had hard times. The forest very badly impacted and the food sources hit hard. We will go there. And then walk inland to see a harvest site. It will be long day."

Shortly thereafter, the group found their hammocks and settled in. The breeze blew gently through the night. The hammocks weren't great but were better than the hard floor.

They boarded the smaller boats at first light after a quick breakfast of rice. Niles had read that jaguars are fabled to roam the forests beyond the riverbanks, although rarely seen. Giant otters had long called the river home, although recent overfishing greatly reduced their populations.

Several "extractive reserves" are situated along the

Purus. The intent was to provide local communities with approved use of the land, mostly for rubber extraction, to stop the cutting of trees for soybean, thereby allowing a more traditional, sustainable use of the forest. Mostly families and small rural producers operate in these areas, but illegal logging operations are present, as well.

They passed their first village within a few hours. Children hid behind trees and laughed as the boat passed. What had once been a large alluvial floodplain and vast, wide river was brown and had narrowed in parts, a contrast to the broad Solimoes.

The Purus took them further into the Amazon. They were close to the village of Ipiranga, which was about a four-hour boat ride from Beruri. From there, they turned and went up the Lago do Ipiranga where the canopy closed above them, signaling that they were now deep in the jungle. They had arrived at their destination for the day. From here they would travel by foot.

They stepped off the boat and were greeted by a teenage boy who wore a frown and looked less than pleased to see them. Looking around, it appeared that most homes were communal and set back a ways from the river.

An ancient woman began down the well-worn dirt path towards the river and spoke to them in a language none of them understood. They discovered that the boy spoke Portuguese and could translate. As the crew disembarked, Ricardo offered a crate of canned goods and a large box of crayons, paper, pencils, and other supplies as a gift to the community.

Ricardo explained to Niles and David that the people here spoke a native language, known as Yanomami, but some of the younger villagers were exposed to Portuguese or Spanish from the traders passing through.

The group walked slowly behind the woman. Ricardo followed the teenager off to the left to deliver the crates. David, Niles, and Paulo were joined by two younger boys who began to follow them past the dwellings ahead. The woman pointed with her walking stick to a small outdoor gathering area just beyond where they stood. They turned their attention to a large clearing that overlooked the river.

Thatch strapped to log poles provided shade. Logs served as stools. Niles had been sitting most of the day and decided to stand. He looked at Paulo, who seemed uneasy. "I am not sure where our contact is. I had expected him to meet us at the boat," he said in a hushed voice as the two smaller boys looked on with speculation.

"Well, I need to find a tree to water," Niles said.

"Be careful. The snakes here, one bite, it is goodbye," Paulo said with a smile.

Niles walked about 30 feet down the path and stepped off onto a smaller dirt path that went off to the right. It looked recently used. Things grow back in the jungle quickly, clearing paths is a continuous process. He looked over his shoulder to make sure no one was around and turned back towards the tree.

The next thing he knew, the world went black.

Paulo heard them coming. Twigs snapped nearby. He hollered for David, who had gone down closer to the water, and pulled his gun. David stopped in his track and reached behind his back for his gun just as three appeared out of the woods, each loaded with their own gun. The villagers were nowhere to be seen, hopefully hiding in their homes.

At that point, David and Paulo realized Niles was missing and Ricardo had gone to the other side of the village to drop off the goods. It was two versus three.

Paulo was the first to take aim. One clean shot took down the lead man. His weapon dropped as he crumbled to the ground. Bullets sprayed from the gun of the second guy. David soon realized that he was out in the open with nowhere to go. He dropped to the ground, took as much cover as he could, and fired off multiple rounds. The second man stumbled to the ground. Then David noticed that Paulo was down. He rolled over and continued to fire.

Then the shooting stopped. David looked up and saw Ricardo holding the third guy at gunpoint. Paulo moaned. David jumped up and ran to support Ricardo. He undid his tactical rope bracelet and quickly tied up the third guy. They would need him for information.

David went to tend to Paulo. He had been shot multiple times. Most had hit his bulletproof vest but one got him in the leg. A quick assessment proved it was nonfatal and David took off his belt to make a tourniquet. Next, they needed to locate Niles.

"Ricardo, get this guy to talk. What did they do with Niles? He's gone," David said while tightening the belt around Paulo's leg.

Ricardo began the interrogation as an old woman came out to help tend to Paulo. David placed his hands together in prayer and slowly bowed towards the woman as a thank you and then ran down the path Niles was last seen. It had rained recently, and so there were clear boot prints in the mud. But they quickly stopped at what looked like a game trail. They couldn't have gone far. The real question was what they wanted.

They spent the next 15 minutes searching but to no avail. No broken branches, no additional boot prints, no evidence as to where Niles had vanished. They had to deal with Paulo and get reinforcements here to help with the search.

BACK IN KENTUCKY

I walked across campus and saw an athletically built Latina woman walking straight toward me with a very serious look on her face.

"Mimi, we need to talk," she said in an authoritative way that almost stopped me in my tracks. "Come with me."

"Do I know you? I don't make it a habit of following strangers," I said and started to slowly back away.

"I know your father. He is in trouble. I am here to look out for you. Now, please, follow me." She pulled back her coat, which exposed a badge on her hip.

I quickly fell in line and followed her down the hill, past the student center, and to the parking lot out back. She opened the car door and as I slid in, looking up to see my mom across the seat from me. My heart dropped.

"What is going on here? Mom, why are you here? I tried calling you this morning." She looked over at me with tears in her eyes. My mom never cries. Ever.

The woman got into the driver's seat and we pulled out.

"I don't understand. Why are you here? Who is "we?" What happened to my dad? Is he okay?" I started to cry, too. My brain swirled, my palms started sweating, and my heart beating into my ears so loudly I couldn't hear.

"I know you have a lot of questions. I hope to answer most of them when we get there," she said. I leaned over into my mom and she pulled me close. She didn't speak, just cried.

I have no idea how long we drove. It could have been five minutes or five hours.

Somehow, I ended up on a couch in an unfamiliar living room lit only by a few lamps. A filtered grey light streamed in from outside as the sun hid behind rain-laden clouds. A woman and a man talked at a small wooden table in the kitchen. The woman pulled a purple-colored high-back armchair towards the couch. A fine layer of dust coated the dull coffee table in front of us.

"I know you probably have a lot of questions, Mimi, so let me try to answer them. I have already shared all of this with your mother. My name is Anne Rodriguez. I work for the FBI and have been assigned to the same case that your father and David have been working on. I am the Special Agent in Charge here. Your father traveled to Brazil this week to gather information for our case. We learned today that he went missing during a gunfight in a small village deep in the Amazon outside of Manaus," Anne explained slowly and softly. "David and other agents are there on the ground. We are going to find your father."

It was then I realized that there was someone else in the house.

"Hank, can you please bring Mrs. Martin and Mimi some tea and water." She turned back to look at me.

I wasn't able to form any words. I looked at my

mother, who shriveled into the side arm of the couch, her head resting in her hand. My strong mother with perfect posture and composure at all times seemed so tiny. I look behind her at the wall. Generic art. What is this place we are in? How is this even happening? I was just with my dad last weekend, sitting at the kitchen table, having brunch with him and Mom.

Two complete strangers stood in front of me. setting drinks on the dull table in front of me. The woman squatted next to the couch and looked me straight in the eye.

"Mimi, we will get your father back." She nodded across the table.

"This is Hank. He is also working this case with us. The TV here has Netflix if you need to pass the time. We will either be in the kitchen or down the hall in the office if you need anything."

I noticed her tight forearm muscles as she reached for my hand. Her deep brown eyes looked straight into mine. I felt a sense of trust with her. Her confidence and demeanor helped me calm down a little. I reached for the tea and tried to combat the awful dryness in my mouth.

"Thank you," I whispered. She squeezed my arm then she and Hank went into the kitchen where they had a laptop set up.

I watched her for a moment. It was clear she was the one in charge. Her athletic build and tightly cropped dark brown hair made her look the part of a special agent. I like strong women and for a moment my mind was distracted watching her lead these two men.

I took another sip of my tea and slid down towards my mom who lifted herself forward to the edge of the couch to grab her own cup of tea. Her beautiful Korean eyes were nearly swollen shut. I have never seen her like this and it made me realize I needed to be strong for her.

"Hey, are you still watching that show you were always talking about a few months back? What was it called? *The Devil*, or something like that?" Mom asked.

"Oh, you mean *Lucifer*. Yes, I am." Talking about this seemed so normal like our world was not crashing down around us. And then I proceeded to talk mindlessly about the show's premise.

I continued, "I am milking every episode for what it is worth. After Fox stupidly canceled it, the fans saved it and Netflix picked it up, so who knows how long it will last. Those writers and actors are brilliant. I know the whole sex, drugs, and rock n'roll is not really your type of show, but the writing is really good, Mom. A great story arc so far. I think you would appreciate that."

"Do you want to watch an episode together? I have never seen it. Maybe it will help us pass a little bit of time until we get more information from Anne."

"Okay," I said as I stood up and collected the remote off of the dusty TV stand across.

We watched an episode together, although I wasn't able to pay much attention. My mind raced as it tried to process everything. I was just in class, walking across campus, and then suddenly a strange woman took me to a strange house, and here I am with my mother, waiting to see if my father can get rescued from somewhere super far away. I feel completely helpless. I heard a voice and realized my mother was talking to me.

"Okay, so let me get this straight: The devil is on vacation in LA, decides he wants to be a detective, falls in love with a detective but won't allow himself to believe he is in love with her and she is also possibly in love with him but won't admit it. And then she turns on him when she realizes he is really the devil, which he has been telling her this whole time, but she did not believe him. And he has an angel brother on earth, and God is playing games with all of them." She paused to process it all. "An interesting take for sure. I think my favorite character is the one that does the forensics."

"Oh, that is Ella Lopez. She is amazing, isn't she?" I said. "And, the crazy thing is, my new roommate, Kiki, is kind of like her. She is some sort of crazy tree forensics person. She develops all kinds of tests for telling where trees really come from based on their genetics."

And then I remembered that Kiki was on a plane headed to Brazil for her conference and field trip this week. I make a note to mention this to Anne.

"Well, now that is interesting. I look forward to meeting her soon," Mom said.

"Mom, Kiki is on her way to Brazil right now. I hope she is safe," I said, now more worried than ever.

My phone buzzed and I about jumped out of my skin as I fumbled for it, hoping it is my father telling me he is safe. But it was Damian. I had not heard from him in a while and had begun to think he wasn't interested anymore. I didn't have time to deal with him right now.

Anne heard the ping. "Mimi, I am going to need you to turn your phone completely off for a little bit."

"Hey, have you seen Meems today?" Clara asked Carter as she walked into the kitchen. He and Nik were having a deep discussion at the table. "Oh, sorry to interrupt. I am just worried. It is not like her not to return my texts and normally we would have met at The Bean for a coffee between classes but she wasn't there today."

"No, I heard her talking to Sherp this morning saying she could not go climbing today because she had too much to do for school. She is probably just off studying somewhere," Nik said and turned back towards Carter.

Clara checked her phone again, no response. She went to the fridge to make herself dinner.

"Hey, Carter, I am starting to get a little worried. Mimi's not like this. When she says she is going to meet me at The Bean, she shows up," Clara said.

"I bet she is with that guy from the Club she has been all hot for. Don't worry about it. Friends flake sometimes," said Nik.

Clara pulled out her phone and pulled up Damian's number. She saved it just in case she needed to pass it along. She decided to text him to see if he had heard from her.

The phone buzzed back, *Hi. No. We haven't talked in a few days. Is she okay?*

"Well, she is not with Damian," Clara announced to Nik and Carter. "Now I am worried."

"Well, where does she normally study? I would check there first. If she's not there, maybe check other normal hangouts. I would not freak out yet," Carter said calmly and casually.

"Not Mimi. She never flakes out on anything. She is the kind of person that thinks being on time means you should be early and if you are actually on time then you are really late. This is so strange." Clara paused and thought for a few moments. "But, good idea, I will head over to the library and The Bean and see if I can find her. Thanks," Clara said as she wrapped her peanut butter and jelly sandwich in a cloth napkin and headed up the steps.

Clara ran up to her room to grab her purse and tried calling Mimi again. Straight to voicemail, not even a ring. Clara flew down the stairs and basically ran to the library.

"Chinese or Thai food? Do you have a preference?" Hank asked my mom and me.

He was such an average-looking guy. If I saw him on the street, I'm not even sure I would be able to pick him out. "Thai," my mom responded with a small smile. "That is Mimi's favorite."

"Yes, thanks," I said as I thought about how Hank could look like any other white guy. I wondered if that was actually a strength in this line of work, the ability to blend in.

"I know how hard this must be. Thank you for being patient. We need you to keep your cellphones off to ensure no one will track you here. We set up a dummy line where we have forwarded your calls. That way, if one comes in from Niles, we'll be able to get it," Anne said as she motioned us to the kitchen. I had no idea how long I'd been sitting on the couch.

"Um, where is the restroom?" I asked.

"Last door on the right." Anne pointed down the hallway.

From down the hall, I could hear my mom and Anne talking. Anne was telling my mom that there was a lead and that they didn't think it was terrorists, but just some sort of local rebellion. As I walked back in Anne was saying to mom, "The United States does not negotiate with terrorists, so this is a good thing. If it ends up becoming some sort of negotiation, we may have things that we can use as leverage."

Anne flipped a laptop around on the kitchen table and showed us a map of Brazil. She zoomed in to the town where my father was last seen. It really was in the middle of nowhere. A sea of green.

"That does not look like the type of place where cellphones work," I said.

"No, not at all. Our agents all carry satellite phones, which catch signals from space, so we have a way to be in touch," Anne said.

"We think the people who ambushed the village thought your father worked for IBAMA. Another

IBAMA agent who was supposed to have arrived at the village a few days earlier has also gone missing. We think someone is just trying to send a message to back off. There are a lot of people in Brazil wanting this area to open up for extraction and development. And there are also turf wars with rival gangs who control mining and illegal logging."

"Hae, Mimi, you need to know that Niles is one of the best investigators I have ever worked with. His tenacity and expertise have supported several multi-country efforts to bring down human trafficking operations."

I looked at my mom. "I thought Dad was a lawyer."

"Yes, honey, he was. And while in DC he also did some other work. But he realized how dangerous this work was and requested to be, let's just say, discharged from duty. That is when he resigned from the Justice Department and moved us to Louisville to start his own law practice. He wanted to be a father and a husband and not bring the stress of this type of work on our family."

Anne paused, stood, and walked to the coffee pot to refill her mug. "He found stuff that no one else finds. His ability to draw out witnesses, with his calm and trusting demeanor, was a real gift. When David and I started working this case, we began to hit some dead ends. David reached out to Niles to see if he could help."

My head was spinning. My dad is potentially being held hostage. I couldn't believe this was happening to my family.

Hank opened the door and the smell of Thai food

filled the house.

TAKEDOWN CITY

David and Ricardo loaded Paulo into the boat and moved as fast as they could back to Beruri where there was a hospital and IBAMA reinforcements. They had called ahead with a satellite phone. Paulo was stable and they were able to stop the bleeding. He had a poultice of local herbs matted to his leg under the bandage and the boat was equipped with a major medical kit. But he still needed medical attention. Another team was divided between two boats headed towards them from Beruri at the same time, they would swap passengers midway, handing Paulo off to the medical team and then the rest would head back to the village.

The ambushers were thought to have two vital assets, Niles and the member of the IBAMA advance team who had come this way a few days ago. The new team shared that customs agents in Manaus had just seized a boat full of illegal timber on its way to the Madeira River. It had likely come from the same operation that the IBAMA agent was looking into down on the Purus.

David used the satellite phone to call Anne. "We have expanded the team and are ready to follow some tracks in the area. We have two local guides working with us. A helicopter is on the way and should be there by the time we arrive to do aerial surveillance. As you know, with the thick canopy, it is nearly impossible to see anything on the ground from the air."

Anne responded, "The family is here and safe. Mimi mentioned that roommate of hers, Kiki, whose research we have been looking at, is in Brazil, traveling in the same area for a conference. The intercepted shipment at the port is thought to have come from the same area where you all were working. She may be an asset to trace the origin of the wood, and possibly a lead to where Niles and the Brazilian agent are being held."

"I will see if we can get someone to bring in Kiki. Most of the small-time illegal logging is targeted at a few specific species and does not go very far. Usually they try to hide a few logs in with some others or process them on-site. Depending on what types of wood we are dealing with, we will be able to tell if these are our guys or not." David walked towards the back of the boat to Ricardo.

"We need someone from the local team in Manaus to find a civilian who can help us with the case. She is here for a research conference tour." David transmitted the info to Ricardo, who then used the satellite phone to make a call to his people.

The boats met up and conducted the passenger exchange, picking up four more agents, armed and in bulletproof vests with large bags of weapons, and a machete for each person. It looked like they were planning to head deep into the jungle. Paulo was taken onto the other boat, which headed back to Beruri. The newly formed team split into the remaining two boats and headed back upstream towards the village.

THEY RETURN

As promised, the mob guys returned, this time at his apartment. The two thugs were waiting for Damian when he left the house for to work.

"Get in," the large tattooed one said. "We are going for a little drive."

"And if I don't?" said Damian. The fat one slid his jacket back to expose a gun.

Damian was unarmed. There was a gun in his apartment but would be of no use to him right now. And he couldn't run. They knew where he lived and where he worked. He had to go with them.

He slid into the backseat of the car, the fat man sliding in behind him. The gun was aimed at Damian's abdomen as they sat next to each other. The other thug, who was covered in tattoos, climbed into the driver's seat. They drove out towards New Circle Road and then blindfolded Damian and zip-tied his hands.

He had no idea where they were taking him or for what reason. If there was one thing that Damian was confident in, it was his tolerance for pain and his ability to defend himself as he'd spent years learning jiu-jitsu from his father.

But Damian couldn't help but wonder what his brother had done to get into this mess. He was a good kid growing up. He mixed with the wrong crowd every now and then, but generally was a good kid.

Their mother had disappeared in Mexico years ago when they were young. At one point, Benji had gone looking for her. Their father said she would come back, that she just was off on an adventure. As they got older, they learned that she disappeared while on a walk one day, shortly after the family had moved to Mexico. Their father and local authorities had tirelessly looked for her. And then Benji continued to look for her. Damian wondered if in his looking, he somehow got mixed up with the wrong people or his heart was so broken and jaded that he just could not recover.

Damian and Benji had been close when they were young but drifted apart in high school. Benji left home at 17, when Damian was 15. Contact was minimal other than seeing each other on occasional holidays and at their father's funeral. Their father had died of a heart attack when Damian was 18. About five years ago, Damian tried to talk to his brother about his path in life, to reconnect and help him see there were options, that he could come to Miami and they could work together. Benji, who had never liked being told what to do, walked out on Damian. They hadn't spoken since.

Even now, all these years later, here was Damian trying to clean up after Benji. When he got the call that his brother had left him a nightclub, he was actually surprised and almost happy. That meant his brother had settled down and chosen to do something with his life, that he had started his own business. Little did Damian know what his brother was really up to. The big question was why. But it didn't matter now. Damian was blindfolded, zip tied, and in the back of a car with a gun pointed at his side.

The car stopped. It sounded like they were on dirt as he heard crunching outside the window. Maybe gravel? A large hand grabbed his elbow and yanked him out of the car. The air smelled like pine trees and rain soon to come. They were definitely in the woods somewhere.

His blindfold was removed. He was right; They were pulled off of a dirt road in the woods. A black Audi was parked in front of them. A man in a business suit stepped out of the car and approached Damian, who was still held by the elbow.

"And you must be our new business partner," the suited man said. Damian detected a Middle Eastern accent.

"I am not anyone's partner. I inherited this Club from my brother and now I am trying to sell it. If you want a partner, find someone to buy it." The anger in his voice seethed through his teeth. His entire body pulsed with rage.

"Your brother was a good partner. It is a shame it did not work out for him. Such promise with that one. But, you see, that Club is important to me. And so, you will take our money, you will bring in much cash income at the bar and the door, and you will put on music festivals, where the cost of everything is a little bit higher than expected. Are you following me?" The man stepped closer ad his cologne burned Damian's nose.

"The music festival . . . That is all part of this plan, too?" Damian could not believe he had been so stupid to not see this coming.

"It is simple, you help us with our books and we

leave you alone. You get paid a very nice amount for your effort and, as long as the books are clean, there won't be any trouble. You have dealt with money before. I have looked into you. You know what you are doing." The suited man was within striking distance of his foot, but Damian was worthless without his hands.

"And what if I say no? I want no part of this. Take the Club. It is yours. I do not want it. I will go back to Miami as if none of this ever happened." Blood pulsed in his ears.

The fat man threw Damian to the ground. "That will not do. Know this: If you run, we will find you. You already know too much." The suit walked back to the car. "It will be nice doing business with you."

Damian felt the first kick to his stomach, followed by one to the head. Warm blood ran down his face and the unmistakable salty taste filled his mouth. Then he lost consciousness.

He woke up lying on the floor just inside the front door of his apartment. His hands were still tied, and he stumbled as he tried to get up. His vision was blurry and his eye was swollen shut. He shuffled over to one knee, pain radiating down his side as he pushed up against the wall to stand. Limping to the kitchen, he opened a drawer and found a knife. The world was coming into focus now.

He started to remember what happened. This beating was a message. His mouth was full of dried blood. He spat into the sink and checked his teeth. All still intact. He started the shower and sat down in it. The water stung as it washed across his cuts.

He cleaned himself up the best he could and grabbed a bag of frozen peas to put on his eye while he called Tom at the bar to let him know he would not be in tonight. It was the first night that he would not be at the bar since he came to town.

He ate a large handful of ibuprofen and lay on the couch as he tried to process what had happened. He realized that maybe what his brother was into was not by choice, but by force. As he laid there, he replayed the events of the day. He remembered the car, a black Audi. The suit. The smell of the cologne.

That was his last thought before he drifted into a dreamless sleep.

THE SAFE HOUSE

I had fallen asleep on the couch. Another episode of *Lucifer* played on the screen. I woke up, realizing it was not all a bad dream and I was still in this awful house.

"Mimi, love, I am glad you got some rest," my mother said from the purple chair across the room. She was writing in a notebook.

"It has been a long night. The longest of my life. Do you want some tea?" My mother, in her same jeans and green sweater, stood up and walked to the kitchen.

I nodded at her as I pushed myself up and stretched, finding my way down the hall to the bathroom. I passed the office on the way there. Hank and Anne were staring at a network of computers and monitors. Anne looked over her shoulder at me as I passed. She approached me when I came out of the bathroom.

"Hae, Mimi, we have finished our assessment of the situation. We will continue to track the ground team in Brazil. The reports are less frequent given their location and the challenge of communication. I assure you that we have the best people on the job there and we will be in touch if anything at all develops," Anne spoke in a calm, quiet voice that was very reassuring.

"I would like to stay here with you in case Niles contact you," Hae said and looked at me. "Mimi, I think it would be best if you went home. Go to school and do your normal things. It does us no good to have two of us here worrying. Plus, you have finals. If anything at all happens, I will call you."

"Mom, why don't you come to my house at least to shower. You can borrow some of my clothes," my voice cracked as I talked. I suddenly realized how thirsty I was.

She looked at me and nodded.

"If you are ready, I can take you there now," Hank said.

It was about 7 a.m. when mom and I entered the house. No one else seemed to be awake yet, so we walked quietly upstairs to my room. I let my mother have the first shower. I guess Clara heard the bathroom door lock and came running.

"Meems, oh my god. Where have you been? I am so glad you are okay." She flung herself at me. Her wild, curly hair bounced around my face as she hugged me fiercely. She heard the shower and gave me a sly look. "Oh, crap, sorry, you have someone over." She smiled at me.

But I did not smile back. She saw the pain on my face. "Meems, what is going on?"

"Clara, something terrible has happened. It's my dad. On his trip to Brazil, his group was attacked." Hot tears streamed down my face. "He was taken, and we don't know where he is or if he is alive or hurt."

The sobs ravaged my body. She sobbed with me and rubbed my back. When I looked up, Carter was standing in the doorway.

"Is everything okay? I heard you all down the hall," he said with an uncertain voice. The water was still running in the shower. I was sure my mom was in there crying, too.

"My dad is in trouble somewhere deep in the Amazon, and there is nothing I can do about it but wait." Carter walked towards me and knelt by the bed where Clara and I were sitting.

"Oh, Mimi, I am manifesting all of the love in the universe that I can for you." He placed a hand on my knee and looked me in my red swollen eyes. "Your father is a good person, and the universe has a way of looking out for those types of people. With all of the good he has done; I am sure it will come back to him. Have faith." I felt a tingle of hope radiate from my knee to my heart. "Do you want coffee or anything?"

"Yes, please. And some tea for my mom, if there is any."

Carter squeezed my knee and walked out of the room then turned to say, "You can use my bathroom if you want. The shower is really nice."

I realized how much I needed and wanted a shower, and to give my mom some privacy. I laid out an outfit for her and some of my favorite face lotion, grabbed a towel and a clean pair of clothes, and picked up my bathroom caddy then headed to Carter's room. Clara gave me another hug and went back to her room.

Carter was right, his shower was incredible. It had one of those rain-style showerheads. I stood under it until the water turned cold.

I was in a daze as I walked downstairs and saw Carter, Clara, Marcel, and Nik at the kitchen table. My mom was still upstairs. They all looked at me. Marcel stood, walked over and hugged me. Normally, I hated hugs. But right now, they were exactly what I needed.

I had cried myself straight. There were no more tears to shed. I was in defense mode. I was going to help find my father if it was the last thing I did. I had no idea how, but I would.

I heard the stairs creek behind me and turned to see my mother. She had not met all of my roommates before. I quickly introduced them all (minus Sherp, as he was out climbing). Clara followed us to the kitchen. Carter had tea ready for her and coffee for me.

"Honey, I am going to go back to Anne. I saw Hank was still sitting out front, waiting for me." She sipped her tea. "The minute we hear anything at all, I will call you. For now, please focus on school the best you can. Call me if you need anything. And be careful, please. Anne said they would have someone looking out for you. I love you." She took one more sip of her tea and walked towards the front door, grabbing her large purse where she had stuffed her dirty clothes as she left.

I watched her climb into the car with Hank. I saw him say something to her and they both looked at me. There was a car parked across the street. Mom and Hank looked over at the man sitting in it. He nodded, and Hank nodded back, and they pulled out.

Walking back to the kitchen, I felt like I wasn't even in my body. Like I was just floating, an observer of this life. I needed sleep. I was supposed to have a 9 a.m. class, but there was no way I could do that. Not after all that had happened.

I stopped halfway to the kitchen and turned around,

walking upstairs to my room. I locked the door, something I almost never do. My down comforter felt like a cloud.

I was woken by the sound of my stomach grumbling. I had not eaten since the Thai food last night. I fumbled around the nightstand looking for my phone but it wasn't there. I sat straight up.

Where is my phone? I jumped out of bed, grabbed my purse, and let out a deep sigh. It was still in there, still turned off. I was so tired that I didn't even think to turn it back on. What if my mom had tried to call? I internally berated myself, that crazy voice in my head flagellating myself for being so stupid. I turned it on. The phone immediately vibrated over and over again. Dozens of missed texts and calls from Clara, Marcel, Carter—all checking on me. As I scrolled down, I saw the text from Damian yesterday morning.

It seemed like an eternity ago that we had visited the Chocolate Café and had that crazy sex-filled night. I get butterflies, even now in my depression, just thinking about it. I texted him back. *Sorry it took so long to respond. Not ghosting you. Just crazy. Can't put into text. Meet at The Bean this afternoon?*

Shower round two was needed. I had sweated terribly during that nap. I must have been having nightmares. Maybe I could make it to my 2 p.m. class. I loved that class and poli-sci was just what I needed to take my mind off of everything.

The smell of something sweet caught my attention as I walked downstairs. I was starving. I walked into the kitchen and saw stacks of cookies on two plates. Carter was baking. He saw me and smiled. "I was just making some grilled cheese while the cookies baked. Do you want one?"

"That would be wonderful. Can I have one of these, too?" I said as I lifted up what looked to be a cranberry oatmeal cookie. It was not really a question, as I took a bite before he could answer. I slid out a barstool at the prep counter.

"If you need to talk, Meems, I am always here." His back was to me as he flipped the sandwiches. Seriously, what guy is this nice?

My phone vibrated in my pocket. I prayed that it was my mom with some sort of information, but I saw Damian's name. *Can't today, hopefully soon.*

Okay, so that was it. He really wasn't interested anymore. It's fine. I should have known. Someone that hot would not be into me. I had too much on my mind to care about him right now anyhow. It was fun while it lasted. Internally I licked my wounds.

I was brought back to reality by a plate put on the counter in front of me. Carter sat on the stool next to me. He smelled like patchouli, the same stuff Clara wears sometimes. The universal smell of hippie culture.

I found myself telling Carter everything. And he just listened. He did not offer advice and wasn't weird. He just listened. I realized that this could end two ways: My father could either come home or not. I had to believe he would make it home. Carter was right about what he said earlier. My father had manifested so much good in this world, had helped so many people. Karma, or whatever it is called, had to be real. He was the light in the darkness for many. Hopefully, that light would come back to him now.

My phone vibrated again. I looked down. It was my mom. *No news yet. I love you.*

"I better get going to class. Thanks for lunch, and for listening, and for the cookies," I said as I grabbed another one off the plate. Mm, chocolate chip.

"Any time, Meems. And I mean that." He reached for my hand and pulled it to his heart. "This is the love I have, and I am putting it into the universe that it may find your father and bring him home to you."

I felt my insides ignite. I looked into his eyes and he looked straight into mine. I leaned into his body. He enveloped me in a hug and stroked my hair. The warmth of his chest felt safe.

I wasn't sure how long I stood there leaning into him, Carter holding me. Normally, I would have made fun of him with all of his universe stuff, but right now, it was just what I needed to hear. I wanted to believe it.

I withdrew from the hug. "Thanks." I turned and walked towards the door, grabbing my backpack off the rack on my way out.

AS FATE WOULD HAVE IT

Kiki had arrived earlier that day in Manaus for her pre-conference field trip. She brimmed with excitement for the three-day boat tour that would launch tomorrow. They planned to explore parts of the Anavilhanas archipêlego, visit national parks, see waterfalls, and see the unique place where rivers meet but the brown of the Amazon River and dark waters of the Rio Negro refuse to mix, creating a unique parallel flow. She would be joined by ten other scientists and local organizers. Kiki had a personal goal to try to find new samples for her collection and database. She also wanted to talk to some locals about unique trees and high-value trees, maybe even get some DNA and cores. But today was simply time to sightsee in Manaus.

Kiki met up with a one of the other tour participants and one of the locals who worked for the National Institute for Amazonian Research during her check-in. Their local colleague offered to take them on a walking tour of the old city.

During the tour, he pointed out the Teatro Amazonas, which was the oldest building in Manaus, built in 1896 in the Renaissance style. It had a pink façade and a dome painted as the Brazilian flag atop a stunning opera house. The humid climate took its toll on infrastructure, but this building fared well. They made the 15-minute walk across town to the Palácio Rio

Negro, another interesting piece of architecture in the middle of the jungle, with its yellow façade. It was a museum now.

"I guess I am surprised at how urban Manaus is. I had no idea there would be so much culture and interesting architecture," said Kiki as she peeled her shirt away from her belly to air out the sweat. They soon arrived at the hotel and stepped into the lobby, blissed to be in air conditioning. She and her colleagues parted ways.

Kiki headed to the elevator where she was suddenly approached by a man who looked to be a local. "Kiki Johnson?" he asked.

"Yes, that is me," she responded, assuming this was another colleague joining them for the tour.

"I need you to trust me right now and come with me. We need your help identifying some trees. I work with IBAMA and am helping on a case. It is pretty complicated. We have some equipment for you to use, but I am not sure we have everything you need. Are you willing to help?" he said in less of a question and more of a statement.

Kiki was caught so off guard that she had no idea how to respond. "I don't even know your name. How do you know about my work? Is this related to the conference?"

"This is a very sensitive case. If you come with me, I can tell you more. And my name is José," he said as he nodded towards the door to the hotel.

Kiki followed to the door and stopped. "Wait. I have some of my equipment upstairs. Let me go grab it, in case I need it. I will meet you back here in five minutes."

On her way up the stairs, she realized she should probably let someone know what was going on, just in case something strange happened. International texts and calls were expensive, but worth it in this situation. She pulled out her phone and texted Carter, *In Manaus, trip good. Heading to test some trees with some guy named José from an organization called IBAMA. Just wanted someone to know where I was.*

She grabbed her backpack kit from her room, slid her laptop into it, and headed back down the stairs. José stood waiting by the door.

He opened to the Jeep door for her, closed it behind her, and then climbed into the driver's seat. The car had markings showing it was legitimately from IBAMA. The organization keeps a small, unmarked set of offices in Manaus in the industrial district. After their main location on the Madeira River was burned down by miners a few years ago, they have moved under the radar in this area.

On the way to the offices, he filled Kiki in on the fact that illegal logs had been detained just downstream where the Madeira River meets the Amazon River. There were headed to Itacoatiara to be moved. They needed Kiki to figure out what ipê this was and where it came from.

Ipê is a general term that refers to a whole variety of different tree species known for their density and hardness. Some of the species of trees are known for being biodiversity havens, and also for providing Pau D'Arco, which is used to treat all sorts of ailments. There are over 150 species of trees identified as ipê.

José told Kiki that, "While many *Handroanthus* species, an ipê, are growing in areas that are now being certified as sustainable to log here in Brazil, much of what is coming from the forest is not sustainably grown or harvested. It is *overharvested,* in a bad way. So, the poachers, they harvest illegally here, and then hope to mix their illegal trees with legal and certified ones downstream in the market. This wood is worth a lot of money. So much that people are willing to kill for it."

He explained that a little over a decade ago most of the illegal logging trade was done through selling tags assigned to legal logs or through forgeries. At that point, IBAMA was also ripe with corruption. That has changed a lot. But now the power to regulate is with state agencies in Brazil, and there is a lot of corruption happening there with virtual credits for legal logs being sold to mask illegal logs. A lot of money is made on the black market for these credits.

As they neared their destination, Jose told Kiki that they had to have this unmarked set of offices in Manaus in the industrial district because their old main location on the Madeira River was burned down by miners a few years ago.

They walked into the offices and Kiki saw a laboratory through a plexiglass window. There were stacks of tree cookies on the table.

"We have only just begun to collect DNA from trees here over the last several years. So, our database is very small," José said as he showed her into the laboratory. "I hope your database may be more useful."

"My database is open source. We crowdsource the sample information from all over the world. It is a massive global effort for the common good. Non-profits, universities, researchers, and all kinds of people contribute. You would be surprised how much is in here. If you added in your information, it would make it even better." Kiki said as she took off her backpack and started to unload her gear. "I can run these samples and see if we get a hit. If it is one of the ipê species, I am sure we have it in the database. If not, at least you will have them for your records, and I hope you will let me add them into the database." She took note of the chemicals on the shelf. It looked like they had everything she would need.

"To run the samples and do the chemical analysis is going to take the rest of the day and a good part of the evening. At least this wood is very fresh. That will make it easier. I wish the trees hadn't been stripped and that there was some bark to work with, too." She grabbed the first cookie within reach and went to work.

"Listen, I have to be in the hotel lobby at 7 a.m. tomorrow for my Amazon adventure. I will help you as much as I can but I am not going to miss that." She flashed José a definitive smile. "Start by telling me about the area where you think these trees are from."

Jose went on to describe the area. "The trees that were approved under a plan to be legally logged in this specific part of the forest were away from the river, adjacent to a coal mine. We think the trees here were harvested much closer to the river and upstream a bit, outside of the approved zone. We had been tracking some activity in that area for the last few weeks after local villagers alerted us that they were threatened by people working near the area."

Kiki ground samples as he spoke. She planned to run several tests, one where she'd put the DNA into the system to see if she could identify the exact species of tree. Then she was also going to run a panel to look at the rations of non-radioactive isotopes of oxygen, hydrogen, and nitrogen. Trees absorb nutrients all the time, which can help tell the story of where they were located.

She sang and hummed as she processed the samples. José brought Kiki some rice and beans for dinner as she diligently worked the samples. Then the first hit came back.

"This tree is actually not in the ipê genre at all. It is *Paubrasilia.* I cannot believe I am personally holding a fresh piece of brazilwood in my hand. This stuff is really rare these days."

She ran samples for several more hours and found that there were several ipê species in the mix. There were also low-value trees like tanimbuca and timborana. She even found one within the *Dipteryx* species but she couldn't get an exact match.

The chemical testing was a bit more sensitive, and more of a forensic science than most people would expect. They had a mass spectrometer, which surprised Kiki given the semi-remote location and lab configuration. She ran samples to determine which trace chemicals and isotopes could be found. She didn't identify any affiliated with mining from the first set.

After taking a quick break, she ran another few samples and found all were from the ipê set.

José told her about some recent busts they had made and how understaffed their team was. Some agents were responsible for patrolling almost 600 square miles or more of dense rainforest territory. Sometimes they would catch loggers in the act, but they more often found stuff at ports like this.

"As much as we try to be proactive, we are mostly reactive. It seems like they are always one step in front of us," he said. "The problem is that after they cut the forest illegally, it is not like they leave, and it grows back. Most of the time, they are cutting it because someone will come in with cows or to mine, so they are making the inroads for permanent forest conversion. That is the real problem. And, not to be strange, but you should try karaoke sometime, your voice is *muito bonito*."

Kiki suddenly became very excited. The last set of the *tanimbuca* and *timborana* trees showed trace minerals associated with mining. She had found a match. Her profuse sweating may have added to her glow as she thought that she may have helped solve the case and found the trees. She cheered out loud, "Gotcha!"

José walked quickly across the room. She smiled at Kiki and looked at the sample. She could tell he was anxious and excited.

"This confirms what I had thought. They mixed the batch. They took low-value species from their legal plots and mixed it in with the illegal trees, the ipê, the *Dipteryx*, the brazilwood. I bet they planned to mislabel the ipê as *tanimbuca*. They do that a lot. The good news is, in recent surveys, we found very few brazilwood trees in that corridor. So, I have a really good idea where it might have come from." José turned to leave the room. "I have to make a quick call. I can take you back to your hotel. I will be back in a few minutes."

Kiki was exhausted. It was very late and she had just spent the last 12 hours at this lab. She ensured all of the sample's data were logged in her computer and began packing up her things. José came back into the room. "Ready?"

"Ready." They walked out into the night back to the Jeep.

"Why are you doing this?" Niles asked in a quiet voice to the man standing over him. His arms were tied behind him to a narrow tree. The man did not look at him. Niles assumed he did not speak any English. He tried Spanish instead, *"¿Qué quieres ganar haciendo esto?"*

The man looked at Niles, but still did not respond. Niles head pounded from where he had been struck. He tried to assess his situation. It was hard to know what time it really was. There were six men that he could see. Daylight filtered through the canopy, so they were somewhere in the forest but not near a clearing. There were a few tarps strung about and a fire pit in the middle of their space. There was some sort of forest bird on a spit over the fire, indicating that this was not a long-term established camp. Niles noticed a line of fish hanging between two trees, so maybe they were still close to the river. They were definitely still near the village, as he had been unconscious and would been difficult to carry a long distance. He tasted the musty smell of decay in the forest and took note of the vibrancy of the different greens around him. Tree frogs chirped close by.

As far as he could tell, he was not injured in any other way. They had not beaten him, which meant he probably was not going to be pumped for information. He was likely being held hostage for some sort of ransom. Out here, the only way to communicate was by satellite phone and those do not work under thick forest canopies. If these guys were communicating with someone, they had to be going to a clearing somewhere.

Niles then remembered that he had an IBAMA satellite phone in the backpack. He tried to remember if he had the backpack with him when he walked off to use the restroom. He was pretty sure he had set it down by the stoop where David was standing. Satellite phones are difficult to track, but do emit signals. It is possible that it could provide a clue to his team as to where he was.

Niles attempted to make small talk with his guard. His Spanish wasn't great, and his Portuguese even worse, so he went with Spanish. *"¿Eres del pueblo local?"*

"Cala a sua boca!" (Be quiet!) the guard said in a grumbling voice.

"Realmente disfruté de este bosque. Es bastante hermoso," Niles struggled to find the right words to say he was

enjoying the beautiful forest as he tried to make small talk.

"Eu sou do brasil, aqui nos falamos português," (I'm from Brazil, here we speak Portuguese.) the guard snapped. But that did not deter Niles. He knew that the languages were close enough and that the guard had clearly understood him earlier.

"Está tratando de conseguir dinero para mí, sé a quién llamar. Sé quién pagará," Niles was trying his best to say that if they wanted money, he knew who to call. The guard walked over and kicked him hard in the side. There was for sure a steel-toed boot under the muck boots. Niles grunted and leaned over, doing his best to coddle his ribs.

"Eu ja falei pra você calar a sua boca." (I said to be quiet.) The guard walked up to the two men standing near the fire spit and talked to them in hushed tones. But Niles knew he had gotten through to the guard.

Niles was almost certain that these men were not part of some well-organized crime syndicate, but rather locals who were hired to do some logging or mining in the area for big payouts.

Niles heard two voices in a heated conversation behind him in Portuguese, and then one voice shifted to English and said loudly, "We need to get out of here. These guys have no idea what they're doing!"

He heard a loud crack and a moan. Niles realized that the man behind him must have been the IBAMA employee who had gone missing the day earlier.

The other voice came from around the tree and faced Niles, standing above him looking down. Niles noticed he was very short and thin. He carried a machete around his waist and a gun on the other side. His clothes were dirty, wet, and ragged.

"Americano, você vale muito dinheiro," he said. (You are worth a lot of money, American.)

"I can make a call," Niles said, giving up on his broken Spanish and Portuguese.

"You here, *por quê*?" the man said in broken English mixed with Portuguese, seemingly hoping to determine Nile's intention in visiting the forest.

"Para ver la selva tropical y conocer gente local para un estudio de investigación," (To see the rainforest and meet the locals for research.) Niles said.

"Eres un cientifico?" (You are a scientist?)

"Sí." (Yes.)

The men walked away and started to speak in their native tongue. While these people likely had little to no education and hardly any resources to back them up, they were certainly well-armed.

For the rest of the day, the men came and went from the camp. They provided scant amounts of water and even smaller portions of food. As night fell, the two captives were moved closer to camp. They were now close enough to talk to each other.

When the men walked away, Niles heard the IBAMA officer say, "They are scared and mostly harmless. They do not seem to have a plan. I think the one has access to a satellite phone, but I am not sure. It is not yet clear what they want or why we are here." Niles was happy to hear they may have taken his satellite phone, if it indeed was his.

He didn't sleep much that night. He spent most of the night feeling things crawl on top of him and wondering what snake might find its way into his lap. He also assessed how to get out of the ropes and which captor he would target first.

The dawn sunlight crept through the trees. A man with the gun walked back towards the Niles and the IBAMA guy.

"Eu te passo esse telefone, e assim você liga pedindo dinheiro," (I'm going to give you this phone so you can call for the money) he said.

It was indeed ransom that they wanted. Niles nodded yes. The guy walked over to the tree. Niles knew that they wouldn't be able to get a signal in the camp. He would be untied and brought to a clearing.

"No señal aquí, necesito apertura, abierto," (There is no signal here. I need space, an opening.) Niles said as he looked up at the sky.

The man signaled to one of the others to come untie him. From here, Niles' words and actions would need to be carefully calculated.

They began walking away from camp, the leader and two other men on either side of Niles. The other stayed back at camp. Niles asked them if they wanted money for the IBAMA guy, too.

"Não, queremos que o IBAMA pare de nos incomodar." (No, we want the IBAMA to leave us alone.) The leader said, indicating the IBAMA guy was being held for a totally different reason, they were sending a message for IBAMA to stop getting in their way.

Just as they started to enter the clearing, Niles heard

the distinct sound of a helicopter in the distance. It was hard to hear anything under the thick canopy of the forest with all of the insects and birds chattering incessantly and the howler monkeys screeching.

All three men also looked up and stopped just short of the clearing. They had only been walking for about 10 minutes, so they were still close to camp. Niles did a quick visual sweep of the area to see if he could break free and run. As he did, he noticed two men standing in the forest on the other side of the clearing. He remained completely still, held only by the arm by one of the men.

Within seconds, they were surrounded. Niles intuitively dropped belly first to the ground as his captors were seized by the Brazilian authorities. They were gagged and brought to the ground. Niles stood up, nodded at the man standing next to him dressed in a ballistic vest and full camo. He briefed them on the situation back at camp, letting them know they were holding the IBAMA employee as well.

Niles quietly led the group back to camp, carefully retracing his steps. Niles pointed ahead and remained behind as the group of six officers fanned out. Two officers remained with Niles and turned to walk him back the way they came. Then he heard gunshots in the distance.

The helicopter landed in the clearing as Niles and the two officers approached.

"How did you find us?" he inquired.

"We had a general radius from the IBAMA sat phone ping, and a scientist visiting in Manaus had run samples of a seized shipment and identified a special brazilwood tree, which we knew had to come from this area. The helicopter spotted the smoke, but we did not want to spook the bastards who took you. So, we sent a ground team in by boat." The officer nodded towards the helicopter. Niles headed towards it. "Let's get you checked out. Do you need water?"

About five minutes later, Niles saw David and Ricardo come through the clearing. He stood to greet them. They were followed by two men in handcuffs, who were gathered up with the other three. Niles also saw the IBAMA agent. He was glad to see that they were all safe and in one piece.

David smiled at Niles. "Good to see you, old friend. Let's get a call into Anne so she can let Hae and Mimi know that you are okay. We will all convene back in Beruri in a few hours for a debrief."

Niles and the IBAMA agent climbed into the helicopter and headed back up the river.

RANSOMLESS

I was startled awake by a knock at my bedroom door and sat up in bed as it opened. It was my mom. She came running over to the bed, crying and smiling. "They found your father. He is fine. Perfectly fine. Barely a scratch."

I wasn't sure how long we sat there, but the catharsis was real. I was so relieved. The last day and a half had been awful. My mom told me that some local kidnappers were just trying to get ransom, and that David and the ground team had rescued them easily.

Mimi looked at the clock. It was 8 a.m.

"Let me get up and dressed, Mom. Feel free to help yourself to some tea in the kitchen." I climbed out of bed and headed to the bathroom.

When I came down, my mom was sitting at the counter in the kitchen with Carter standing across from her and Clara sitting next to her. She had shared the good news with them already. I could tell Clara had been crying and I knew what was coming. She jumped up and ran over and threw her arms around me. I hugged her back. Carter walked over and hugged me, whispering in my ear, "The universe provides." That made me smile.

"I already made you coffee. It's next to your mom." He smiled that Carter smile at me. We were all catching up when my phone vibrated with a new text. I slid it out of my back pocket and glanced at it under the counter. I was surprised to see that it was Damian. I thought that train had left the Station.

Sorry I have not been in touch. Club stuff. Want to meet at The Bean?

I put the phone down and looked at my mom. "When can I talk to Dad?"

"He should be back in Beruri soon. He said he would call as soon as there was reliable service. Do you want to come with me over to see Anne? She may have more information by now?"

I nodded.

Clara looked at us both. "My heart is so happy right now. I love you all so much."

My mom gave Clara a big hug and nodded at Carter as we turned and left the kitchen.

Back at the safe house, Hank met us at the door and led us back to the room where he and Anne had their computer set up. I looked at Anne. Even though I knew she had been up all night, she was still stunning and on point. She stood as she saw me and walked over. Her hand reached for my shoulder.

"I am glad you came and that we had good news to share. Your father was lucky," Anne said, her dark brown eyes looking into mine.

"Back here, we have been communicating with the ground team in Beruri. Your father's helicopter arrived safely and he is being evaluated by the medical team. I am expecting we should hear from him very soon. There is coffee in the kitchen if you want some."

I was overcome with emotion. Hearing this news

from her made it more real. "Thank you. I am okay. I just had some. But I might get some water." I walked into the kitchen.

Anne was on the phone when I came back to the office. She handed the phone to my mom, who she started talking in a soft voice to who I assumed was my father.

"Is it Dad? Put it on speaker, please."

She nodded, said something else to him, then clicked the speaker button and placed it on the small table in between us.

"Dad, oh my god, I am so glad you are okay," I said, tears also falling down my face.

"Oh, honey, I am fine. After those two days in the jungle, my father figure is more of a dad bod. The food was terrible out there." I knew he was okay if he was back to his terrible dad jokes.

"I should be back at home by Sunday or Monday, if the weather and flights cooperate. We just have a few more things to finish up here before I head home. I am really looking forward to seeing my two favorite girls."

"I love you, Dad, and cannot wait to see you," I said. A wave of relief washed over me.

"Well, there is no way that I would miss my favorite daughter's graduation," he said.

It was a standing joke of his that I was his favorite daughter, it was because I was his only daughter. Then reality finally hit me again. Graduation was coming up very soon, like in two weeks.

Mom took Dad off of speaker and walked out of the room.

Anne looked over at me. "Mimi, this part of the case is done, but we still need to be careful. I am going to have our agent continue to keep an eye on your house for now. I don't want you to worry, just to be diligent, okay?"

"Okay," I said, "but you need to know, I don't want my roommates to get in trouble with the police. They smoke a lot of pot."

She chuckled. "Oh, we noticed. And no, we are not worried about that at all."

I walked out to the living room, sat in the purple chair, and pulled out my phone as I waited for Mom. A few minutes later, I saw her hang up and walk back into the room to talk with Anne and Hank.

I texted Damian back. I knew I needed to spend the rest of the day studying, but my head was totally in the clouds right now and I couldn't focus on studying even if I tried. Plus, I wanted to see what he had been up to, and if we still had a connection.

I am up for The Bean, how about 1pm?

My mom walked back into the living room and smiled at me.

"Are you going to be okay? I am thinking I will go back home today, since everything seems to be okay with your dad. I can stay in town, though, if that would make you feel better."

"I have to study for finals, anyway. I am good."

"Okay, honey. I will see if Hank can run you home." She got up to walk out of the room and turned looked at me. "Your show, *Lucifer*. I may have watched a few episodes last night. I still cannot believe Chloe betrayed him like that, talking to that priest."

That is so out of character for my mom. She never watches television. She reads and writes and listens to classical music. Her head must have really been spinning for her to watch one of my shows.

"You're hooked, aren't you," and I smiled at her and we both shared a laugh that was desperately needed.

My phone buzzed and I looked down to see a response from Damian. *Looking forward to it. See you at 1.*

I wasn't sure if my heart could take any more emotion, but I was going to try.

Hank had dropped me off at home. It was lunchtime so I scrounged the fridge for some food.

Marcel walked in. I couldn't remember the last time I'd seen him, it seemed like ages ago, but was probably only a few days.

"Meems, I am so sorry. I heard what happened to your dad. I'm glad he's okay." I knew another hug was on its way to me. For a person who was not a hugger, I had given and received a lot of hugs lately, it was almost starting to feel normal and not terribly awkward.

"Thanks, Marcel, I just talked to him. He is doing good and hopes to be home in a few days. Is there anything to eat in here? I am starving."

"I am making a PB&J if you want one. The ol' standby. Clara and I are upstairs working on our final stories. I was bringing her one, too. I can't believe that in a little over two weeks we are done, like done. DONE. With college, forever." He smiled as he pulled out the gigantic jar of peanut butter from the cupboard. I went and grabbed the bread and a few plates.

"I had totally forgotten about graduation with everything going on."

"I guess we first have to pass our finals." He laughed as he slathered jelly onto the bread.

"Yeah, I may be coming back to school next year to bring my GPA back up. I have not been able to study at all. I guess we'll see, right?" I grabbed a bag of chips from the snack drawer.

"You will be great, Meems. You know this stuff; you studied all semester. You should read Clara's story so far. It's really incredible. She's going to make an awesome journalist. See you at family dinner tonight?" he said as he picked up the two plates. "Clara and I are cooking. It will be good."

"Wouldn't miss it." I bit into my gooey sandwich and promptly finished it in about 1 minute, I guess I was hungrier than I realized. When I am stressed, I really like to eat crunchy things. Potato chips were my weakness. Salty and crunchy, you cannot beat the combo.

Ten minutes later, I was out the door to meet Damian. My mind told me that this was a mistake, that I should stay home and study, but my heart was telling me to go. I promised myself I would be gone for no more than an hour. I brought all my stuff with the intention to head to the library after to study.

The Bean was packed with a line out the door. I

walked up and scanned the line for Damian but did not see him, so I queued up. I felt him before I saw him. My body went on high alert. My stomach tingled. I looked over my shoulder. He was about two feet away.

"Hey, you," he said, smiling at me. I could have melted at the sound of his voice. And then I noticed the huge bruises on the side of his face and the slight crack on his lip.

"Damian, oh my god. What happened to you? Are you okay?" I instinctively reached up for his face. He flinched.

"Oh, this? Yeah, I am fine, just a little mix up at the Club. That's all." I could tell that he was trying to play something off.

"Well, it looks like more than a little mix up." I eyed him carefully.

The line moved quickly. We ordered our drinks and then grabbed an open table outside under the big oak tree.

"So, are you going to tell me what *really* happened? Is this why you couldn't meet up the other night?"

"Meems. . ." He reached across the table for my hand. "There is some stuff happening that is outside of my control right now. My brother was involved in something. I am not sure what yet. I may not be able to be around much until I figure it out. I want you to be safe, and I am not sure being with me is safe right now." I saw the sadness in his eyes. He was being genuine.

"Safe? Ha, I am being followed by police right now. I haven't exactly told you what is going on with me. Long story short, my dad was captured and held hostage in Brazil and was just freed this morning and the police think that might be related to something going on here. So, if you are worried about me, don't be. I have protection." I looked around to see if I could find anyone watching me.

"What? Wait. Your dad was being held hostage? Why didn't you say anything? That is crazy, Meems. I can't believe you have been going through that." He leaned across the table and brushed my hair behind my ear. "I am very sorry I wasn't there for you."

I sat back. I couldn't handle his touch right now. It was just too much. I realized this whole thing was too much. I needed to go study and focus on school. I have a little over a week before finals.

"It sounds like we have both been going through some crazy times," he said as he looked at me. "Meems, I want to protect you right now. I am pulling back because I am not sure how this is going to play out. I don't want to get you mixed up in something that is no good. I know you feel what I feel between us. I don't want to step back, but for now, I would suggest staying away from the Club." He seemed so incredibly sad.

Heat rose in my face. It was anger. I was on the emotional rollercoaster right now. The Club was one of my favorite places and he was telling me I shouldn't go there. I was going to crazy town in my head, so decided to take a few deep breaths and pause. What if what he was saying was true?

"It's fine, Damian. I have a lot to process right now anyhow. Finals soon, then graduation. My best friend was shot at, my roommate was shot, my dad was held captive by who knows who in Brazil, I probably failed my second LSAT . . . I have no idea what I want to do with my life. It is probably better to have some space right now." I stood up and pushed in my chair.

He grabbed my hand. I looked down and saw his forearm tattoo and then let my gaze follow the length of his arm up to his face, his cut lip, the faded green bruises around his eye, and then met his eyes.

"I am not giving up on you, Mimi Martin. Take care of yourself and be careful." He stood up, pulled me close, and kissed me. I stood there for what seemed like an eternity then stepped back and walked away towards the library. The rest of the night was a sorry excuse for studying.

WHERE THE LIVESTOCK SLEEP

Pacing back and forth in the last stall, the sound of hay and dirt crunching under the rich leather boots, Damian startled from a sound above and looked up to the rafters. Sweeping wood ceilings, inlays framing the entrance, the horse barn looked more like a multi-million dollar mountain lodge than a place where livestock sleep.

He heard voices outside the barn and froze, hunkering down behind stall door. He didn't recognize the voices.

"I can't help that it was seized. But do not worry, it was only one part of the order. The other part is coming by boat. You should be getting what you need by the end of the week." The voice sounded like was laden with years of cigarette smoke and a heavy life.

"If I don't, you know there will be consequences," the other voice said. It sounded like he had a Middle Eastern accent.

Damian questioned his sanity in coming here. He had years of martial arts training, had done his compulsory 12 months with the Brazilian army, was skilled with handling firearms, spoke four languages. . . Why was he hunkered in a stall stalking potentially super dangerous criminals? Why did he not go straight to the police? His calf started to cramp from squatting for so long.

The voices left the barn, but he was not willing to move yet. Damian had memorized the Audi's license plate and had been following it from a distance for the last two days, tracking it to this farm. It came here at the same time two days in a row, so he made a gamble assuming it would end up here again today. Several cars and trucks came and went from the farm. The property was huge. He noticed that they had cameras on the barns, so he cut the wire on his way in. It was quite possible he had already been caught on camera, but he was pretty sure he was obscured before he cut it. He stowed his motorcycle in the woods about a half mile down the road and hiked to the barn. The farm was surrounded by beautiful rolling hills of green lined with split-rail fences on three sides and a well-managed forest on the west.

After several minutes had passed, he scouted his options for getting out of there. Seeing an opening, he quickly ran back into the forest. He had no idea how this horse farm was connected to his Club and what his brother was up to. Maybe some sort of shipment?

Damian located his motorcycle, pulled it out from the bushes and he rode the winding roads back to town. His mind raced the whole ride.

INFORMANTS

Back in Beruri, the local officials and David had isolated the captured men and were pumping them for information. As one of the more skilled negotiators, Niles asked if he could have a go. The Brazilian officials said that would be fine once they were finished. One of the officials had some limited knowledge of the local language and Portuguese as his native tongue, so he was serving as a translator with David and Niles.

Niles began by telling the man he was doing research on forests and local cultures. He asked about this man's village, what it is like, how many people live there, if they subsist from the forest mostly or where they work to buy goods, the education level of the children, and how they learn.

The man stonewalled him at first. Niles pivoted and told the man about himself. His home was mostly surrounded by forest, too, but a different type. The man appeared to be listening to him.

Niles offered the man water and food. He ravished the plate as if he had not eaten in quite some time. He was still nervous and sweating, not from humidity but from fear. Niles noted this.

"Listen, you can tell me stories about where you live, and we can continue to feed you. But if you do not talk, you will be back with the police. I don't know what they will do to you or where they will take you. It's good for you to talk to me right now," said Niles as he slowly bobbed his head up and down, trying to get the man to acknowledge he understood.

The man bobbed his head back. Another plate of rice and beans was pushed across the table to him. One of his hands was cuffed to the table, the other free to eat with a plastic spoon. The rural prisons in Brazil are notoriously bad. People are treated horribly and regularly die. Niles knew this from his experience representing Americans who had been held in a Brazil prison for minor infractions. Their treatments were horrible. He had graphic images still engrained in his mind from their stories.

"So, do you live in a village on the Purus River or did you come to this area from some other village?" he spoke slowly so the translator could repeat.

"Purus River. Close to camp. Small village. Five families," was what came through the guard serving as translator.

"How many children do you have?"

"Seven."

"Do you rely on food from traders or from the forest?"

"Mostly forest. But kids want more from traders. They come more now. It is harder. The forest is going away in places, harder to find food. Harder to stay away from outside."

"You said the forest is going away. Can you tell me more?"

"The animals know when the men come, and they run. The water is not so clean and the fish harder to get."

"When do the men come?"

The man was starting to sweat more. He remained silent. Niles could tell he was weighing his options.

"If I tell you about the men, can I go home to my family?" As the officer translated this, Niles looked at David and the lead officer standing in the corner, making eye contact with both.

"Let's not get ahead of ourselves. I am just trying to learn more about your village. Can you tell me if the men you were with are also from your village or were they from other villages?"

The man stopped talking. He had finished the rice and beans. The guard took away his spoon and cuffed his other hand to the table. Niles could see that the Brazilian officers were getting annoyed with this storytelling approach. They had a more direct approach to getting information. While this man had not killed anyone, his group had shot one IBAMA employee and had opened fire on a whole squad in the forest. The Brazilian officers were seething resentment.

Niles stood up and walked out of the room. David and the Brazilian officer followed.

"I think I can get him to talk. Can we offer him his freedom if he does? We all know he is likely a dead man walking if we cannot figure out who was directing and paying him before he gets back there."

The Brazilian officer made some calls. He soon returned and agreed that they could negotiate, but the price of his freedom was dependent on the information he'd provide. Niles walked back into the stark room. It was grey. Paint peeled from the ceiling and off of most of the walls. The cinder block construction was obvious. Built to last, not to look good.

Beruri did not have a prison, but it did have a sheriff's office, the *Delegacia Interativa,* in the heart of town, which is where they were sitting. It was a simple, brown roofed building at the intersection of two dirt roads. They tended to the issues of Beruri's approximately 20,000 inhabitants, but also to issues in the surrounding areas. The police were mostly locals. They observed the shakedown with interest and amusement.

Niles sat back down across the table from the man. "Tell me your name. *Meu nome é Niles*." He at least knew this much Portuguese.

The man shook his head. Niles leaned on his elbows on the table. There were two Coca Colas next to him, a real treasure out here. He caught the man looking at them and slid one across the table.

"We will work with you, on your request. If you give us the information we need, you will be able to go back to your family. So, let's try this again. My name is Niles. I live in the United States. I am studying trees and villages and local people here. That is what I do. Now, please tell me about you."

"Many people have many names where I come from. You can call me Mukashe. I am from the village upstream." The translator spoke.

Niles had conducted extensive research on the Purus River. It ran for nearly 2,000 miles from Peru through the Amazon, where it finally meets up with the Amazon River in Manaus. Along its meandering path are vast wetlands and several small villages that mostly rely on fishing. The river, as he had seen it, was a vibrant orange-brown color, wide and with good volume. He had also studied most of the transportation options between Rio Blanco and Manaus, developing an understanding of where extractions could most easily occur.

The Purus runs through some remote terrain. Until Brazil's 230 road was built in the 1970s and then the 317 road was completed in the early 2000s, most of the river between Boca do Acre and Manaus was inaccessible. It was where some of the remote last tribes lived. In most of these parts of Brazil, you only travel by foot or a small wooden canoe.

"Mukashe, how did you come to Ipiranga?"

He shifted in his seat. "Men come to village with guns and tell us to work or they kill our family. We cut down sacred trees for them and our soul's cry. Not all were sad. They brought us from different places. Some men had not heart or mind. They just wanted money. They paid small money, but we could not leave. They gave us guns and chainsaws and told us what to get. They gave us donkeys to pull the logs to the river. They would bring boats to take the logs."

Niles could tell the man was quite shaken by all of this but did not seem to have ill intent.

"The men, here, they think there is much forest. Enough for them to take and everyone to take and keep taking. They say that it will make Brazil great. The money. But the cows and mines, they make the water bad. And the animals in the forest go away." The prisoner's head bowed. The translator looked at Niles.

"Mukashe, can you tell me more about the men, the ones who came and took you from your village? Anything at all you remember? And how did they get in touch with you to tell you where to go next?"

"One man was in charge. He had the sky phone. He talked to people and then told us what to do. When they came to our village, they acted like traders trying to sell cans of food and candy. They were in a cargo boat, very nice. Only two guys from boat came into village. They offered a ride to the next village, where I needed to go. And that was the last time I saw my family."

The conversation went on a while longer, with the guards becoming increasingly frustrated by the lack of detail that was coming from the man. But Niles remained patient, slowly drawing it out of him. It was clear he was terrified. He had probably never been to a city, never been to a place like this before. Most of the villages had wooden huts built on stilts near the river, ramshackle ramblings cobbled together with supplies from the forest, and maybe some supplies bought from traders. No electricity, no running water. Here, the overhead lights flickered, the airless room stifling. Niles could tell it was overwhelming for the man.

Niles learned that the man was one of three villagers who had been conned. For the other three, one worked for the big man and two were regulars on the logging industry circuit, moving from camp to camp willingly to make money.

Niles realized that they were not going to get any more information from this guy. He really didn't know much. They needed to talk to the guy who worked for the big man. Turns out that he was the one who was shot during the raid. He'd survived but was at the local clinic being treated.

"He has told us what we need to go and held up his end of the deal," Niles said to the guards. "Good luck," he said sincerely as he turned to head out the door to the clinic.

Niles and the others headed to the clinic. When they arrived, the man who had been shot was sitting up in bed, handcuffed to the rail. It was a small clinic with minimal supplies and old equipment that looked like it had been donated from another hospital years ago. One guard was sitting in a chair to the side of the bed, the other leaned out of the open window, taking a drag off of his cigarette. It was clear that the local police were not as interested in this case as the national authorities.

For the next hour, Niles and David interrogated the man in the bed with the help of a police interpreter. The Brazilian officer waited patiently for his turn with the man. The prisoner said very little about who was directing operations and why. He did rail, however, against the environmentalists who were ruining the livelihoods of local people. With roads, there was now finally a chance for Brazil to develop, to begin thriving, and poor areas had resources and could become rich. But stupid laws and rules prevented this. "Forests grow back, we should use them," he kept saying. Niles had run into this type of thinking before and wanted to lecture the man on the difference between sustainable forest management and deforestation.

The prisoner spoke perfect Portuguese, so it was clear he was an outsider and did not have the dialect of the local people. It seemed obvious he was only in this for the money. Niles had an idea and discussed the plan with David and the Brazilian officer.

"We are prepared to offer you safe passage to anywhere in Brazil, along with 100,000 reals if you tell us who you are working for and how the operation works," the Brazilian officer said to the man. The man looked around the hospital, surrounded by guards and police, Americans and Brazilians. He knew he was trapped with no way out.

"I do not know who is in charge, only that we get instructions by satellite phone from a man in Rio. His name is Mohammad. There are many of us that he directs. He gives us goals for what to cut and where to take it. We get more money for special trees, like brazilwood. After we put it on the boat, we do not know what happens to it. There are people at the ports who work for him. They forge the documents or they overlook some trees that have come in. I know they sometimes sell them as logs, and sometimes they get processed into flooring or decking or other things before going out of the area," the man said. He looked over at the guard by the window and asked if he could have a cigarette. The Brazilian officer nodded his head yes and the guard handed one over.

"Would you like to sit in a damp room watching water drip from the ceiling? On good days, you may have a sliver of light reach your skin. Yells echo through the halls in the night. Men have no thought about you other than for their own passion." The Brazilian officer looked scornfully at the prisoner. "What you have said is not enough information for your freedom. To avoid or famous Brazilian prisons, we need you to call him on the satellite phone and provide us when and where to meet him."

HOW DO YOU KNOW HIM?

I looked at my cell and had two missed calls from a random number with no message. Ugh, telemarketers. I walked from the library to my ethics class. Instead of a final exam, we had a final paper, and it was a total nightmare for me. I really liked this professor, Dr. Holmes. I even more so liked the class, Environmental Ethics. But I am still a little embarrassed that I bared part of my soul to her. Fortunately, she acts like it never happened.

I pursued an ethics focus throughout college to provide a strong backing for my law degree, but some of the classes I took just because I was curious. I was pretty much a nerd for this kind of stuff. But Environmental Ethics had my heart. My climate activism passion was fueled in this class. I would find myself feeling honest anger, almost rage, at some of the things I read about.

I decided to write my final paper on environmental justice in underserved communities. Specifically comparing Beattyville, a coal economy turned drug-ravaged town in eastern Kentucky that sits on the North Fork of the Kentucky River, Clinton, a rural agricultural hub in western Kentucky, and Louisville, a big city on the Ohio river in the middle of the state. My brain was so focused on this paper that I didn't see Anne standing there when I got out of class.

"Mimi, do you have a minute?" she asked. She was in street clothes and had on makeup. She looked like she could have been an athletic instructor or a professor. She fit right in.

"Um, yeah, of course. Is everything okay?" My thoughts immediately went back to my father who was still in Brazil. He was supposed to fly home tomorrow.

"Yes, it is. I tried calling you a few times. When you didn't answer, I figured I would just come find you. Let's take a walk." The two of them slipped into a comfortable pace down the pathway that led out towards the library.

"Listen, you know we have someone keeping an eye on you. They shared an interaction with me, and I need to ask you some questions. What can you tell me about Damian Lopez?"

I was caught off guard. Why was she asking me about Damian? They must have seen my body language at the coffee shop yesterday.

"Well, he owns the Club. That is kind of how we met. He is newer to town, moved here from Miami to take over the Club. His brother died and left the Club to him. He is a nice guy." I found myself picking at the lint inside my jeans pocket. Why am I nervous? This is Anne. She does not care if we slept together or not, does she?

"Okay, that is a good start," she said and stopped, pointing to a small wall surrounding a grassy area. There was no one around. We both sat. She crossed her legs and leaned toward me as though she was going to tell me a secret. She had on tiny diamond earring studs.

The sun caught one of them, which amplified it against the dark color of her skin.

"Mimi, why don't you tell me a story? Start from a few weeks back, whenever it was that you met Damian. Tell me everything between then and now that you can remember." She leaned back on one arm with an intent look on her face.

I filled Anne in on everything, starting when I first spotted Damian at The Bean and ending with our conversation yesterday. "I saw him yesterday at The Bean and we broke it off. He said he had some stuff going on that he didn't want me involved with. I guess I should tell you that it looked like he had been beaten up. That is how we left it. I am busy with school, he has some Club stuff going on, and the timing is just not right." It was at this point that I realized how broken my heart felt. I really had feelings for Damian. Like an electric compulsion. But the feeling did not seem deep, more like surface lust and passion.

"That is really good, Mimi. Really good. Thank you for sharing all of those details with me. Did he by chance say what his situation with the Club was or who beat him up?"

"No. No details. We have never really gotten into details, actually," I said and looked at my watch. I now only had about an hour until my next class. I had really hoped to make some progress on the paper between classes.

"I understand you need to get back to class. Thanks for talking with me," Anne said as she stood up. I was starting to cook in the sun anyhow, midday heat in

Kentucky is no joke with the humidity we have here.

"Is Damian okay? I mean, is he in trouble or something?" I asked.

"Mimi, it is important that you don't mention this to him. I think Damian is okay. We have just been seeing some odd behaviors around the Club and were looking into it. It was great to see you. Good luck on finals," Anne smiled at me and stuck out her hand.

We shook and parted ways. I headed towards the library. Great, another distraction to put into my brain. Now Anne is looking at Damian and the Club. At least I know that he was legit. Some strange stuff was definitely happening at the Club.

THE CONFERENCE

Kiki arrived in Curitiba after an incredible Amazon adventure. She and her colleagues had seen the river dolphins, collected samples of all kinds of plants for the database, and had slept in hammocks under the open stars. She was exhausted and exhilarated.

Curitiba is in the southeastern part of Brazil, home to the famous *Araucaria angustifolia* tree and to many forest research facilities that work in the Atlantic Rainforest ecosystem. There was a large conference facility here where the research convention was being held. She had almost two full days of attending the conference and listening to presentations. She was inspired. Her presentation was tomorrow and tonight she and her two co-author colleagues, a woman from Vietnam and a man from China, were meeting to review their slides. She found it very odd that her Chinese colleagues always had the same guy with him, who said nothing, like a sidekick.

They met for an hour in the hotel lobby. When they felt good about the presentation, they met up with the rest of the research group to go out to a traditional Brazilian steakhouse. Many stories were shared about travel from all over the world. One colleague had to walk three miles to catch the bus to take a train to then take three planes to get here. Others had come early and done other pre-conference tours. They all exchanged tales of adventure.

Kiki couldn't hold it in any longer; she had to tell them about how the new technique had actually helped solve a case. In her charmed voice, she articulated the drama that had unfolded, from being approached inside the hotel to working through the night to finally identifying where the trees had come from, and that IBAMA was going to go catch some bad guys. It was like she was a secret ninja for tree science. After too many rounds of meat on sticks, the group walked back across town to the hotel.

The presentation the next day was a huge success. Kiki got a lot of questions, and there was a line of people waiting to speak with her. Some wanted to collaborate, others wanted more detail about the DNA database. She buzzed from the positive feedback. She was taking the red eye home, so she listened to a few more presentations and then decided to sneak out early to go visit the Jardin Botanical Gardens before her flight.

The Jardin Gardens are housed in a beautiful greenhouse situated in the middle of a large meandering park. she noticed a man watching her. This was not entirely strange as many people look at her because she was a sprightly Black woman with blue hair. Still, he gave her an uneasy feeling. She slowly and methodologically made her way through the rest of the greenhouse and noticed it was starting to get dark out. She decided to head back to the hotel and grab her bags then head to the airport. She sang to herself as she headed to the park exit and did not hear the footsteps approach behind her.

Her arm was ripped backwards by a strong calloused hand. His forearm quickly wrapped around her neck and covered her mouth and he dragged her backwards. She was being pulled off into the trees on the side of the park.

Kiki bit down on the assailant's hand that covered her mouth, but it did not matter. It took her mind a minute to rewire into the mental catalog that was her lifetime of martial arts training.

She wrapped her right leg around his, creating a hook behind his knee. She took her free hand and pried at his pinky finger. Once she had a good grip, she threw her head backwards and smashed his nose, at the same time prying his finger around and swinging herself loose as she collapsed his knee.

She managed to get a brief look at him. He wore a mask and a hood. And he was short, maybe only a few inches taller than her. She sprinted out to the main road and got in the first taxi she saw. She shook as she entered the hotel. She went straight to the front desk and asked for her bags. The taxi was still waiting for her, per her request.

Kiki dashed to the airport went with a vengeance.

BACK FROM BRAZIL

When the landing gear grinded underneath the plane, Niles knew the descent was almost complete. He was trading tropical humidity for midwestern humidity, he was not sure which would be worse; but for certain, the bugs in Brazil were worse, much worse. He had itched his bites the entire plane ride home. When the wheels hit the ground, he turned on his phone and texted Hae.

Hi, Love, I have landed, home soon. Cannot wait to see you.

He waited for the row in front of him to clear, grabbed his bag out of the overhead bin, and exited the plane then headed for baggage claim. As he exited the secure area, Hae was there smiling at him.

"You could not have expected me to wait at home for you. I grabbed an Uber here. I am so glad you are okay." Niles and Hae fell into a long embrace.

"Well, I do not miss all this travel. I am for sure retiring, I mean really retiring after this case is over." He grabbed her hand in his left, held his briefcase in his right, and they walked towards the carousels.

"Niles, you were kidnapped in the middle of the Amazon jungle. I was so worried about you. Mimi, she was beside herself. I don't know what we would do without you. I fully support this real retirement you speak of." She smiled at him.

"You are lucky this was a discreet job, or the media would be all over you right now. Lucky for you, not a word of it made it anywhere that I can tell. It was really hard not being able to talk to our friends about this. Mimi shared what was going on to her roommates, so be aware there."

"I am just so glad to be home and with you. I may need to head to Lexington in the next day or so to follow up on a few items, so I will be sure to see Mimi when I do. And then her big graduation is next weekend. I can't believe our baby is graduating." He grabbed his bag and they left the airport for home.

Damian weighed all of his options for how to move forward. There weren't many he could tell about his situation. Selling the Club was not an option. Walking away from the Club and going into hiding *may* be an option. He had not yet determined the reach of the criminals who were threatening him. Taking on the mob himself was not an option. Going about daily life and doing their bidding was against everything he was as a person, so this was definitely not an option.

He needed an ally. He needed help. He needed information. He needed to find someone he could trust.

He called Mimi and was surprised when she picked up, given how they had left things the other day.

"Damian, is everything okay?"

"Yes. Hi, Meems. Everything is okay. I hope it is all

right that I called you. Listen, you said you father was a lawyer at one point. I think I need some advice and help from someone I can trust. Is there any way you would be willing to put me in touch with him? I know that is a lot to ask of you, but I don't know where else to turn right now." He sounded calm yet desperate.

"I can ask him. But he is just getting back from Brazil, and with all that happened, I just don't know what his state of mind is or how much he is working. But let me ask, okay?" Mimi said. She wanted so badly to ask him to come over. He wanted so badly to ask to see her. The tension was palpable.

"Thanks, Meems. I really appreciate it." He hung up.

Back at the house, I sat at the big dining room table and hacked away at my ethics paper. I was almost done and about jumped out of my skin with joy when I heard the snip from *Jellicle Ball* ringtone play. My dad was calling.

"Dad! You're back! Oh, I am so glad."

"Hi, sweetheart. I'm back. You know, I learned something in Brazil. Guess what kind of shoes frogs wear?" He paused, waiting for her to queue him.

"Really, Dad? Okay, fine, what kind?"

"Open toad. Ha."

"Seriously, you just back from being kidnapped and are telling bad frog jokes. I guess I am glad to hear those crazies didn't destroy your sense of humor. So, are you okay? I bet Mom was freaking out when she picked you up."

"Mom is right here and while I did not freak out, I am very happy to have your father home," Mom chimed in. I guess I was on speaker.

"Hey, hon, I have to come to Lexington tomorrow afternoon or the next morning to do a few things. Would really love to see you. Is one of those better for you?"

"Actually, Dad, tomorrow afternoon is great. I can't wait to see you, too. Hey, um, weird thing to ask, but I have a friend who was hoping to talk to you. He needs some advice on something, I think he may have got mixed up with the wrong crowd or something. Any chance you would have time to meet him tomorrow while you are here?"

"A friend? Is this a special friend?" He put the emphasis on "special" and waited for my reaction, which I did not give. "Regardless, if this person is a friend of yours, I would be happy to talk to them."

We finished up the conversation.

"Meems, do you know why Spiderman is such a bad boyfriend," I knew my mom was rolling her eyes just like me. "He is just too clingy," he was cracking himself up. At least one of us was laughing.

"Meems, you can look forward to more of those tomorrow, I love you dear, talk soon," said my mom as we hung up.

I texted Damian. *My dad will be here tomorrow and can meet you. I will let you know a time later.*

The rest of the night flew by. I was reviewing flashcards when a plate was placed down beside me. I looked over and Carter slid onto the bench.

"I figured you could use some fuel. It seems you are working overtime," he said, smiling at me. He had made me a bunch of carrot sticks and homemade hummus.

"Wow, Carter, thanks. That is super thoughtful." I crunched down on one of the carrots.

"So, flashcards. Interesting. I guess I should have taken you for a flashcard girl. That seems your speed." He laughed. I blushed. I had never really thought about any other way to remember stuff besides flash cards, it was just my process.

"Hey now, what is that supposed to mean?" I joked back at him.

"Just that I am impressed with your level of dedication and organization. Whatever you end up choosing to do, I am sure you will excel. I have never met anyone quite like you before." He slid off of the bench and walked back to the kitchen.

I studied until my brain couldn't absorb anything else and decided to go to bed. I placed my law book under my pillow. Maybe osmosis would work somehow.

It was easy to wake up the next morning. I was excited to see my dad. I jumped out of bed and hit campus early to study and finished all of my classes by noon. Dad was treating me to lunch at the Mediterranean place, my favorite. We were set to meet there at 12:30. I am always on time, or early, never late. I get that from my mother. My father, on the other hand . . . While never late for work, he's always late for everything else. Today, however, he showed up on time. He looked a bit haggard, but otherwise all in one piece and no notable bruises or broken bones. I hugged him extra hard to see if he would flinch and he did not.

He grunted as I squeezed him. "Mimi, do you know why the skeleton needed a hug?"

He studied me for my reaction I shrugged my shoulders.

"Because he had no-body!"

"Oh my god, Dad. Seriously, you are so punny. No, really, the bad dad jokes, please make them stop." I laughed as I pushed back from him. We walked to our table.

The curtains of the restaurant were opaque lace. From outside, you couldn't see anything inside during the daytime, except through the small door window that was covered in latex letters that provided the name and phone number and hours of operation, topped by the Jordanian flag. There was very little outside light. Inside was dimly lit, as buffets normally are, and there were tall-backed fake red leather booths lining either wall. With the loud Zajal music, few customers, and even fewer employees, this was a great place to meet for private conversations. But I had really come here for the food. It was authentic and delicious.

Dad and I got caught up on what he was able to share with me about the trip. Basically, that the Amazon was amazing. Alive. Green. Rich in diversity. Mesmerizing. He used that word twice.

"So, Dad, this guy, Damian, my friend who you are meeting with, I am worried he may have gotten himself into something. He looked a little beaten-up last time I saw him. I know he is a good guy. Yes, don't even ask. I *do* like him, but we are just friends? And don't be surprised, he is a little older than me. Don't freak out, okay?" My father just nodded.

"Is he still planning on meeting us here?" he asked as he dipped his pita into what was the absolute best Baba ghanoush in the world, or certainly in Lexington.

At this point, I was oblivious to the concept that my dad might already know who Damian was. In my mind, he was just some Club owner with a great forearm tattoo that the universe placed in my path.

"Yeah, in fact he should be here any minute."

As if right on cue, the door opened and in walked Damian. He walked slowly through the restaurant, looking around as he did not see us at first.

When he got to our booth, I noticed how nervous he looked. At least the swelling in his face had gone down and it looked like he was healing okay.

I looked up at him and smiled. This was super awkward. "Damian, hi, this is my dad, Niles. Dad, this is Damian Lopez, my friend who wanted to talk to you."

I noticed Damian had dressed up a bit for the meeting. Gone was his usual attire. He wore fitted slacks that fit his body perfectly, and a light pink tucked-in button-up that showed off his fit waist and brought out the beautiful olive tones in his skin. He was a masterpiece in this outfit.

"Mr. Martin," He offered his hand. "I really appreciate you making the time to talk with me today. Mimi, thank you so much."

"So, Dad, I am going to head back to the library to study. Text me if you need anything before you leave town. I am so glad you are back and are okay. Damian, it's good to see you." I smiled at them both and slid out of the booth. On my way out, I asked the hostess to turn up the music a little. She obliged.

Damian remained standing until Niles motioned for him to sit down.

"Damian, Mimi tells me you might be in some sort of trouble. Is that what you want to talk about today?

"Mr. Martin, I don't know how else to say this, but I am a good person and I fear that, out of no doing of my own, I am involved with some bad people. I am afraid and I don't know who I can trust or what I should do," Damian spoke in a very low voice, leaning on his elbows across the table.

"Tell me more Damian. How are you involved with these people?"

"I came here when my brother died to take over a Club he left me in his will. I have been here a few months and noticed that his books were off. The other day, some guys, mafia maybe, came in and threatened me, and then roughed me up to send me a message that I worked for them whether I liked it or not."

"And did you go to the police with this information?"

"No, small towns like this, I don't know the politics

or who is who or if there is corruption."

"Damian, do you know what your brother was involved in or who these people are?"

"Not really, no. It seems like they want me to launder money for them, but I am not sure where the money is coming from or who they work for."

The bells on the door jingled. Anne came walking in. She approached their table. Damian was clearly on edge.

"Damian, I want you to meet Agent Rodriguez. She and I work very closely together." Anne dipped her chin as a courtesy nice-to-meet you nod and slid into the booth next to Niles.

"I have not been completely honest with you here. I know who you are, Damian, which is why I agreed to take this meeting. Agent Rodriguez asked my assistance conducting some research for an investigation and, in the process, I came across your Club," Niles explained.

"Damian, I think we can help each other out here, if you are open to that," Anne began and waited for his acknowledgment.

Damian remained silent.

"We need access to the guy running the show; whoever is harassing you at the Club. We were watching a property the other day and noticed a motorcycle just off the road near the property with plates that belonged to you," Anne paused.

Damian continued to remain silent. He weighed if he should get up and leave now or if he was trapped. It appears that he was trapped here by what appeared to be the police and on the other side by a mafia-type organization. He needed to play his cards right.

"Damian, I am sorry if this is catching you off-guard. You came here thinking you were meeting with a lawyer who might provide you with some legal advice or, at least, guidance. And that meeting is also happening. Unfortunately, this is also a situation that you happen to be involved in, without knowing it. If you really are innocent and want out, then please help us. If you help us, we can help you." Niles grabbed his tea glass and took a drink, looking over at Anne.

She sat there resolved, her hands crossed on the table, staring at Damian.

Damian shifted in his seat. "How do I know I can trust either of you, especially after you ambush me like this?"

"This is not an ambush, Damian. This is me bringing in my colleague to show you that you are not alone. We have information, just not *all* of the information we need. You can be the key to getting that. If we win this, then you are free." At this point, Niles was the one leaning across the table.

Anne followed up, "Damian, this does not look very good for you, you know, owning a Club and being at a property under investigation."

Damian slid out of the booth. "Okay, I'm in. But don't go getting me killed. And I don't want to be seen with you all. I am leaving here now in case they followed me. You all need to stay behind or leave a different way. Find a way to get in touch with me that does not involve the Club. And soon." He walked out the front door. The bells chimed behind him.

BIG BREAK NEWS

"Meems, it seems like it has been *forever*!" Clara announced as I walked into the house. "We seriously need to go out. It has been too long. Want to go to the Club tonight and see the Avett Brothers cover band with me? I am sure you are all studied out by now. I need a distraction from this story. I am at a dead-end and all of the words are blurring together."

"Ha. Clara, do you know me at all? I am 100% committed to studying around the clock until finals are over. *Then* I will go out with you," I said as I slid onto the couch next to her.

She sat cross-legged, her laptop propped on her knees. The TV was off. Music drifted in from the kitchen. I heard Marcel say something to Clara and then he saw me. He still has his sling on, still healing.

"Mimi, hey, good to see you. And so happy for you and your dad and family. What a ride these past few weeks have been. I have no idea how you are even attempting to focus on school right now." He walked around to the other side of Clara and sat down.

A plate of cookies was stacked on the table in front of them. Carter must have been at it again.

"Did you tell her the news yet?" Marcel looked at Clara. She shot him an evil eye, which I totally caught.

"No, she didn't. What news?" I looked straight at Clara.

"Well, remember back last fall when I was applying to all of those fellowship programs?"

"Yes. And . . ." I said staring straight at Clara, who was hiding a smile.

"And . . . I was accepted Meems! They picked me. I am moving to New York after graduation! I got the *New York Times* Fellowship! I mean, I must have been the backup to the backup given how long it took them to notify me. But I don't care. I got it!" She beamed with pride, and Marcel beamed right along with her.

"Of course, you did! You are amazing, Clara. I am so incredibly proud of you!" I realized that I was crying. "The stories you did covering that campus scandal with the trust fund misuse, and then the athletics program cover-up, and the fact you were editor of the school paper. You are a rockstar."

She cried with me. Underneath our excitement, we realized that, for the first time in our lives that we can remember, we will be apart. I looked over at Marcel. Clara uncrossed her leg and kicked him. "Tell her."

"I am going, too. I landed a *CNN* diversity fellowship. My first step towards real international reporting." His passion shined through.

Energy radiated from them. Life was working out for them.

"The universe is providing. All of their good energy is coming back to them." Carter walked into the room while taking a swig of his Fat Tire. "Pretty amazing, huh! I can say, 'I knew them when.'"

"Wow, both of you, moving to New York. Are you moving too-geth-ther?" I slowed down for emphasis.

They looked at each other and smiled. I had my answer. "I am so very happy for you both! We do need to celebrate, for sure. Let's have some cookies. Then, after finals, a proper celebration!"

"Agreed!" Clara chimed in. "My final story is mostly done but I don't know how to end it. And I only have 72 hours to finish it up. Long-form."

Carter offered to give it a read, which made me want to offer to read it, too.

"I will take a look, Clara. Not that I will be of much help, but it will be a good diversion from law and economics," I said.

That night, I read her story. It was a wonderful feature piece on Kiki and her new techniques to track down a new-age style of criminal. Before I knew it, I fell asleep thinking about all of this.

EUREKA MOMENT

My brain fired up as I drank my morning coffee. "Clara, don't you think it is kind of crazy to think about all of this illegal wood stuff? I mean, what if what happened to my dad in Brazil is related to what happened to you and Marcel at the mill in some weird, totally unlikely way?"

Over the steam in her mug, Clara scrunched up her nose. "No shit, Sherlock. I know in my gut that they are related. Something crazy was going on at that mill. I just can't figure out what. That is my great conundrum and why it is not in the story. I'm worried about digging too deep, especially after what happened last time." She paused in momentary thought. "I do have an interview this morning with a lutherie that uses rare woods. He is going to tell me more about the industry. At least I can just go with the story on Kiki and the wood trade in general if I can't break some crazy crime syndicate story happening in our backyard."

Geez. Clara was working in real-life stuff. I was over here in fluff, hypothetical land. I can compare Kentucky towns all day long, but is what I am writing about the environmental injustices actually going to change anything? Is anyone actually going to see it besides my teacher? The little existential crisis that started to have a few weeks ago crept back in. Here comes the self-doubt.

"I really, really am happy for you. I can't wait to hear

more. All of it. I want every detail, okay?" I stood from the couch. The old walnut floors creaked beneath my feet. I turn towards our stairwell. My hand grasping the rail supported by the ornate lathe-spun balustrade beneath. The well-worn chestnut stairs cushioning my way up as they had generations before me.

I processed everything that's recently happened. Clara and Marcel, moving to New York. My mind flashed back to the fact Damian and my dad met yesterday. I was dying to know what it was all about, but I know my dad would never tell me and it was clear Damian did not want me to know. I hoped he was able to help Damian.

I closed my bedroom door, sat at my desk, and flipped open the laptop. I was distracted by the beautiful prism of light filtering through the old lead windows that I hung my crystal in front of. Focus, Mimi, focus. Only a little over one week left.

MEANWHILE

Meanwhile, back in Brazil, the meeting had been set. The informant reached out to arrange the next drop. From that, they were able to collect the needed information. They assigned two officers to track the local ringleader for a day and find an opportunity to reach out to him to set up a meeting. He was an easy target. Obviously, he had gotten lazy and comfortable in his role, assuming he was protected by the corruption around him. The Brazilian officials brought in an agent from Sao Paulo who would pose as an interested buyer for a high-end flooring company. From there, they hoped to reach the main node.

David remained in the area and as part of the team. The IBAMA professionals were back to their normal day jobs. David admired them for their perseverance. They weren't paid enough, nor did they have enough protection to do the jobs they were asked to do. Impossible jobs, really.

David and the other two US agents that had joined the team identified any possible links of this operation to the growing US operation. The Brazilian officials were cautious of their presence. They had also been tracking this network but here you never know who you can trust, and who has turned for the cash and ruin an operation with one phone call. The team had to be tight with each person highly vetted.

The meeting was set in Santarém, a port hub for hardwoods, on the Amazon River's path to the Atlantic Ocean. David would accompany the Brazilian officer in the undercover operation to pose as a potential American buyer, a partner in the flooring company. David looked the part. Tall, well-kept, tight white beard against his brown skin, able to look every bit the wealthy businessman when he put on anything other than his field clothes.

The seller wanted to meet near the port docks, Porto da Amizade, on the bank of the Tapajós River, where it meets the Amazon River before traveling the remaining several hundred miles out to the Atlantic. The port was massive, the second largest in Brazil, stacked with containers. It was dated, having been built in the mid-70s, and heavily used.

With the roads built into the Amazon, crossing all four of the regions, Amazonas, Para', Mato Grosso, and Rodonia, the port of Santarém was necessary to move the growing flow of raw materials out of the area. It also drew tourists in cruise ships who wanted to see the green-blue waters of the Tapajós River meet the muddy brown waters of the Amazon River.

David and the officer prepped the local team. The Brazilian would do most of the talking and work on negotiations. David would frequently inquire about specific types of wood products he was looking for.

They parked the car and walked down towards the docks. A cruise ship had just come in and the streets filled with disembarking passengers. They walked down the park-like boardwalk of the Vila Arigó towards their destination.

"Interesting mix of boats," David said to the officer, noting the number of small fishing canoes and boats tied up alongside larger showboats. David also noted the number of soy storage facilities and the large ships with infrastructure built out over the old beaches to move the grains from the storage to the boats.

"These grains, the soy and corn shipments out of this area, have put Brazil on the map as a leading exporter. It has lined many pockets and fed many families. Sometimes you ask at what cost," the officer said as he pointed to large boat stacked with raw logs docked ahead.

That meant they were getting close to their destination, a factory-style building adjacent to the port. They continued down a dirt path past shanties made of corrugated metal where locals pedaled goods. Arriving on foot was easier to be more casual, like they were coming from the fancy tourist hotel. Teams of agents were strategically placed around the area, on standby. They blended into the crowd.

David and his Brazilian partner, Raul, arrived at the warehouse and were greeted by two men in professional casual wear. They were Brazilian and likely the muscle but posed as business associates. Introductions were made. At this point, it seemed like a legitimate business operation with real partners eager to seal a legitamate deal.

The warehouse had large rollup doors on the east side that faced the water. This allowed for large trucks to pull in and out. They also had a large space out back that was a large dirt landing where logs could be stored. The second half of the warehouse was a processing plant. Logs were made into planks and boards here, used mostly for flooring and decking. Many of these trees were legitimate. Some even held certificates of sustainability. This made the work of tracking the illegal ones even harder, but the science of tracking and understanding these illegal systems was improving rapidly. For every one bad actor there were ten good ones, and these guys were making harder for the good ones to keep operating.

The officer, Raul, and David were invited into a conference room that overlooked the factory floor. Large plexiglass windows separated the storage and meeting facilities from the factory operation. A polished man with his pomaded hair and too strong cologne talked about Santarém's economic boom and the benefits brought to the region, that these men were supporting rural lives and bringing wealth to small towns by purchasing directly from them versus through third-party sellers. Plus, they had first choice on the quality wood.

David knew these guys made most of their money by pretending expensive wood was cheap wood, and then sorting it out and selling it back on the market as rare. After about 15 minutes of pleasantries and talk of the local area and how the American liked it and how different it was from Sao Paulo's port, Raul said to the men to get down to business.

They were looking for some very special wood for a wealthy and hard-to-please set of buyers for flooring and custom furniture in their new home. David added that he was looking for something similar, but that his people only want brazilwood and will accept nothing less.

"I hear you all are still able to get brazilwood," he said with a sly smile. "I am willing to pay. I have a large flooring order I would like to make for various other hardwoods."

"Sir, you are mistaken. We do not have brazilwood here. That is illegal to harvest these days. But we do have a very nice selection of other hardwoods, including some that are difficult to get and could be of interest," the polished one said. "Let's go take a look."

The men went down to the factory floor and looked at various hardwoods, selecting several of interest, and then made their way back to the conference room.

Raul spoke with an authoritative tone, "I think we can come to some sort of deal. But I am not seeing exactly everything I want. I would like to speak directly with your boss to see if he can help me."

The two men looked at each other. The shorter one spoke, "That will not be necessary. We can handle all of the transactions. The owner, he is not available."

David noticed the camera in the top right of the room.

"I see you are capable, but yet you tell me you cannot get me everything I want. So, again, I would like to speak with your boss. I think it would be a mistake to not treat such a buyer with the respect of speaking to the owner," Raul said in Portuguese.

"Is that a threat?" the polished one responded.

At this point, David knew the ground team had heard everything and was likely surrounding the place. David had faith in the team and in Raul.

"Not a threat at all. It is just that my partner here and I, for the type of clients we work with, we only do business with the top. I am sure you can appreciate this," Raul said.

"Apologies if you were misguided, Mr. Sanchez, but we are the only ones you will be able to meet with. If that means we will lose your business, then I am very sorry to hear it," the short one said in Portuguese. But David knew enough to understand that the easy route to getting to the leader was not going to be an option.

"Well then, you leave us no choice. Have a good day." David and Raul stood and walked to the door.

As they opened it, they spotted the agents in the factory quietly walking along the wall, unnoticed. They headed for the large opening to the two roll-up doors. Four agents came running in as they exited.

David and Raul were handed guns and stood guard on either side of the large exit as the agents entered. Gunshots and a cacophony of yelling in Portuguese begun. Out of the corner of his eye, David saw a small side door from the factory open and a small man run out towards the shipping containers.

"Raul, we have a loose one. Let's go," David said as they ran after the man.

The container yard was not that large and was bordered on one side by water, so there weren't many options to hide. David and Raul split up, Guns raised, they slowly proceeded to check each cargo stack. The hair on David's neck stood up as he slowly turned, finding a gun pointed at his head. It was another polished Brazilian man. He wore an excessive amount of cologne, which gave him away.

"A buyer," he said in heavily accented English. "Now a dead buyer."

A shot rang out, echoing off of the containers at an ear-splitting decimal.

The man crumpled to the ground. David nodded at the Brazilian field agents behind him, took a deep breath, and they proceeded down the stacks. David's ears rang but he thought he heard someone talking in Chinese. The other agent confirmed that he heard it, too. They turned the corner on the next stack and followed the whispering voice. They surrounded him between two large containers.

Guns drawn, two of them, pointing directly at the man. He slowly raised his hands, cell phone in one, and he said something in Chinese to whoever was on the phone. He then threw the phone to the ground as hard as he could and stepped on it with the heel of his boot. David took him out, face down, and cuffed him.

The Federal Police of Brazil and the Brazilian Intelligence Agency shared an office in Santarém where one of the salesmen from the factory and the Chinese man were brought into separate rooms to be interrogated.

GOODBYE

I am at The Bean, sitting at my favorite table, distracted by everything, but trying to focus. I'm a few exams and a paper away from being done with school. My phone buzzed. I read the text from Damian but decided not to respond. There is just too much going on. Plus, we decided it was better to pump the brakes a bit and have some space.

Having already met the maximum number of refills allowed by The Bean, I get up to grab a glass of water and use the bathroom. When I come back, I see Damian in line. He sees me right away and is clearly surprised to see me. My heart lights up, just like his eyes, but I squash that emotion immediately. I smiled at him and turned to go back to my table. He walked over to me.

"Can I join you?"

"For like five minutes, I really have to focus." I try not to look up at him.

"Okay then." He looks at me with his dark eyes. "Mimi, I think I am in over my head, and I really want you to be careful. I am going to back off after today, no more texting or anything until after I get this all sorted out. It is better for you, I promise," he said.

"How do you know what is good for me?" I quip. "You barely even know me. You don't know that I am perfectly capable of dealing with shit on my own. I can defend myself. I am street smart. You don't get to decide what is 'better for me,'" I realize anger is rising inside of me. This happens when I get stressed, I get angry.

"Meems, listen. I have lost people I care about. You don't know what bad people are capable of. You haven't lived in places where people just disappear. Or where every time you take a walk, you never know if a cartel will be engaging in some senseless killing to prove a point. You have lived this wonderful life of nice houses, college, both of your parents being around for you. I want to keep that prism intact for you. I don't want you to end up broken like me. Trust me when I say, right now, not knowing me is the best thing for you." Damian started to get up.

"Damian, I have been through a lot lately. And I may have lived a privileged life, but you seem to have, as well."

"Mimi, you really know nothing about me. You don't know that my mother disappeared, just poof, gone, when I was just a kid living in Mexico," Damian said, his face darkened, overcome by sadness or anger, I wasn't sure.

'When we were young, my father said she would come back, she just was off on an adventure. As I got older, I gave up hope she would ever come back. Benji never gave up hope, he became bitter. My father and local authorities had tirelessly looked for her," pausing he shifted from one foot to the other, hands resting on the back of the chair where he stood.

"Benji continued to look for her. I don't think his broken heart ever recovered. He left home when he was 17, I was just 15. We rarely talked, seeing each other on occasional holidays, and at our father's funeral. He died of a heart attack when I was 18. About five years ago, I tried to talk to Benji about his path in life, to reconnect and help him see there were more options, that he could come to Miami and they could work together. Benji, one who had never been one who liked to be told what to do, walked out on me and we had not spoken since. And now he is dead. So, yes, Mimi, I am a bit broken. And I do know a thing or two about trying to keep people safe," he raised his gaze to meet mine. I had no words. I felt his sorry.

"Thank you for connecting me with your dad. Goodbye, Mimi. Take care of yourself," and he turned and left.

MAHREEN

It was a hot and dry Thursday in Saudi Arabia, much like every other day in Saudi Arabia. But the stakes were high on this day. The ponies raced and bets were made that were not in paper or coin. The number of heads to trade for the winner was significant. The horse industry was thriving, a result of the oil boom and more money to spend than the elite knew what to do with. The slave trade was also alive and thriving. People became a commodity, as easily traded as the horses.

Mahreen had come to Saudi Arabia on business, to meet with an old friend seeking advice on some oil trades. He had been invited to use their private beach resort villa in Jeddah, on the Red Sea. He joined them for a day before they went home to Riyadh, and they offered him their place for as many days as he wished. Mahreen was well-acquainted with the area, having been numerous times for business over the years. He had several colleagues and acquaintances who frequented this resort, which is known for its pristine beaches, exclusive access, and internationally acclaimed diving and snorkeling. He sat by the pool, looking out at the ocean and sipping his Diet Coke.

Lounging next to him were two men in white thawbs. Mahreen pretended to be interested in his book as one man answered his phone. He had heard that an old acquaintance of an acquaintance was coming here this same weekend and that this same person was possibly tied to human trafficking operations. He was the co-owner of a horse racing club in Riyadh and of several thoroughbred Arabian horses that have studded cup winners.

"No, I told you Brazil is guaranteed. The shipments will go out tomorrow. It is done," he said in a harsh and aggressive voice.

"By the end of the month, the transactions will be complete. Shipping is expected to be 20 days to the port in Florida. 26 days to the port in Hong Kong." He abruptly hung up.

The two men proceeded to discuss the profitability of their investments in corn and soy in Brazil. Mahreen kept his ears open, but nothing was said to allude to trading anything other than agriculture and standard horse racing talk.

His afternoon at the coffeehouse offered few other insights. Mahreen was headed back to Emirates in the morning as he had some business there to tend to. His vacation was over. He finished his drink while watching the backgammon game at the table next to him.

"You there, would you like to challenge the winner of this match?" the man said, looking at him.

"It would be my great honor to attempt such a challenge," Mahreen said as he bowed his head to the man.

"It is a game of skill and a game of luck, but much more, it is a game of national pride, and so I should hope to defend my country's honor," the man said as the person opposite him stood to move from the table.

Mahreen grabbed his qahwa and stood to join the game. Saudis were known for their casual approach and unrivaled hospitality.

Before sitting, he conducted the standard introductions, extending his hand to the men before him.

"And so, what is it that brings you to our fine country this weekend?" asked the senior of the two, who introduced himself as Hassan.

"I am only here to relax on a holiday after visiting with old friends. Taking in a little football and a little family time."

"Emirates may have won this last match, but our long history of wins against you will not be forgotten." The man smiled.

Small talk occurred throughout the entire game.

"Your English is perfect, my friend. You must have studied there?"

"In fact, I did. Georgetown University. Spent six years in the United States before returning home. It was an interesting cultural experience. Your English is also impeccable. You must have also studied there?"

"No, but I have much business in the United States. I spend a lot of time in Kentucky and Florida for my work in the equine industry. There is nothing quite like a good horse race to get your heart pumping."

"I also enjoy the races. I have been to Keeneland in Kentucky before. It is a very beautiful place, indeed. So much greener than here." Both men laughed.

"My cousin owns a horse farm in Kentucky near Keeneland. I like to visit there each year. It is truly a magical place."

"Yes, but not quite as magical as this majestic Red Sea and beaches that you are so lucky to have." Mahreen swooped his hand out over the vista.

Mahreen had a chance to win the backgammon game but purposefully allowed his host to reign champion.

"Thank you for the game and hospitality. May your horses all be fast and may your football team be one goal short against Emirates the next time we meet." With a small bow, Mahreen exited the area.

He called Niles and with a brief exchange and shared what he had learned.

UNDESIRED COOPERATION

"Okay, so let's review the plan again," Anne said, looking at Damian.

Damian continued to stew on the ambush that had occurred at the restaurant when they unexpectedly recruited him for this assignment. They had been up for coffee or lunch in various locations for the last several days. To any average bystander, it would have appeared they were dating or old friends. They made an attractive Latino couple. Anne was maybe 10 years older than Damian, but what woman would not be interested in him?

In reality, he was explaining in detail the three men who had come into the Club and disrupted his life. He described the "request" for assistance for the music festival planning and purchasing. He talked about his afternoon hiding in the horse barn listening to two still unknown voices talking about shipments. Shipments of what, Damian did not know.

It sometimes blew Damian's mind that he had gone from managing a start-up to being an FBI asset and was seen as beholden to some sort of smuggling operation for which he wanted no part. He had been beaten and threatened and mostly ambushed into cooperation. He regretted ever coming to this town.

Anne had been in communication with David, who was still in Brazil. Another team member had traveled to Cincinnati to track down another lead. It was at the crossroads of I-71 and I-75 and there was a lot of activity in that area.

In their last conversation, David had conveyed that the man they captured in Santarém was a Chinese national operating a general timber sales company that was owned by a larger shell corporation registered in Panama. They were going to attempt to identify the real owner of the company. For now, the five men that were arrested, three Brazilians and two Chinese, were not talking. Claiming they knew nothing, that they just sold flooring to rich Americans and rich Brazilians. Anyone who would pay. They claimed that they had paperwork for all of the wood and could prove it was legitimate. The Brazilian officials were not buying it; They wanted information on who owned the company and weren't willing to give up until they got something.

After several weeks undercover as a university professor, David determined there weren't enough leads and that he would not return to that role after Brazil. David let Anne know he was going to stay in Brazil for another day and then would stop in Panama on the way back.

Anne was covering the local front and was cutting Niles loose. As a civilian, he had done above and beyond his part. Plus, Niles had just been held hostage, which is more than she expected to happen. Back in DC, there was a whole team of analysts supporting the overall operation. Another team was working with Homeland Security and the State Department in Asia focusing on what had happened in Vietnam. This was a global problem. Anne was now focused on the local leads that Damian was about to provide.

Damian had agreed to continue to launder money through the Club, not just to prevent another beating, but to help the FBI get closer to the source. At this point, he was not sure where the funds were coming from or why the money was dirty. He only knew he was to make some modest improvements to the Club and then help with a local music festival this summer, which apparently was a money launderers haven. No real way to track how much anything costs; Paying carneys to set up rides and tents; Paying random people to work parking. Most of all, each band required cash payments, a super easy way to make money flow out and then back in.

If nothing else, him helping with this festival would hopefully uncover a money laundering racquet. He and Anne continued to chat as if they were friends just catching up at the café.

"And so, you will get in, you will tell them you are willing to work with them for the right price, you have seen the books and see what kind of money you can make," Anne whispered under her breath while taking a sip of coffee.

Damian nodded back in her direction with his water glass raised to his lips.

Today, he would go to the bar and open it like every other day. Except today he would assume a persona that was not him at all. He wondered if this is what also happened to his brother. As he inched his metal chair back on the chipped concrete patio, he put down a $10 bill, tipped the baseball cap that he was wearing, and walked away a new person. He wanted his life back. In order to get it back, he had to make this work.

Anne finished her coffee while studying every person at the restaurant. Twelve people. Two couples, a family of four, a young family with a baby, and two women alone diligently typing on their laptops with earbuds in. Not one of them paid any attention to anyone else. She looked out to the street, clear. She does this everywhere she goes. She studies the people, the surroundings. It is part of her DNA now.

She handed the server enough money to pay the bill and left, heading to the long route around the outside of town and through two unnecessary neighborhoods before arriving at the house.

Damian went home to change. He looked at his phone numerous times, wanting terribly to call or text someone, anyone. He realized how alone he was. No family. No friends. He had been so absorbed in his work in Miami that he had let most of his relationships fall by the wayside. And how here he was, different situation, same feeling.

The band at the Club tonight was supposed to draw a big crowd. Damian showered and dressed, planning to arrive early to ensure the stocking was complete. Seven o'clock rolls around and the crowd filters in slowly at first, then the line builds. For Club owners, this is what you want. You want people driving by to see the line, wondering what is going on, and want to be part of it. He was almost caught up in this feeling last month, the excitement of it. That was gone now, replaced by a sense of dread. He was inside of his head trying to figure out his next few moves when he heard a familiar voice holler at him from the other end of the bar. Large wooden earrings and curly hair bounced up and down as she smiled at him. Clara. He smiled back, nodded, and said something to the bartender and an IPA was handed to Clara free of charge. She smiled and danced her way through the crowd over to his end of the bar to thank him.

"Hey, stranger. How ya been?" Her voice was three times louder than normal.

"Clara, good to see you. It's going. You know . . ." He pushed glasses on the rinse spout and placed them under the bar.

"Well, I tried to get Mimi to come out tonight. It is always more fun with her here, but she is being a good student and studying. Thanks for the beer!" She turned; her arms floated like waves to the music as she moved seamlessly through the crowd. They moved out of her way like they felt the pulse of her energy coming. It was a sight to watch. She must not know about he and Mimi. Otherwise, why would she have brought her up? He was surprised at how mature Mimi seemed. *How could a college girl keep all of this to herself?* he thought.

He saw the tattoos out of the corner of his eye. On a night like this, with this kind of band, it was not the type of crowd to draw scary mafia men. The message boys were back. Only this time, Anne was here, too. She was perched on a stool in the dark corner of the bar, where Mimi normally sat.

The clunky thug demanded a beer from the

bartender while he surveyed the situation. Damian watched him from the other end of the bar. Finally, his gaze met Damian's. He just stared at him. Damian nodded acknowledgment. The message was sent. They were watching him. He turned to pick up his beer, drank it in one go, slammed it on the bar, and turned to leave.

TALES OF A SPACE PROSTITUTE

I stayed home from the Club. Not only had I been told to stay away, but I needed to get in some last-minute writing and studying. I wanted to end the semester on a high note. At this point, I was no longer studying for Harvard or law school; I was studying to get away from all the crazy stuff that happened this month. Cramming can be my happy place. It's like reading a good book where you forget about everything and are right in the moment.

I sat at our big kitchen table, working my way through my brightly colored flash cards when I heard the sound of guitar playing upstairs. It was very soothing. John Denver, one of my dad's favorites. I needed a break, anyhow, so went up the well-worn wooden stairs through the turret. I had planned to sit on the stairs and listen, but the door was cracked open, so I nudged it a little further and Carter looked up.

"Mind if I listen for a few?" I asked. He nodded to the chair next to the door and smiled. He looked straight at me as he sang "You're Gonna Make Me Lonesome When You Go," by Bob Dylan.

I should've felt awkward or uncomfortable, but this was Carter. He isn't my type at all, and I'm sure I'm not his. He probably liked those girls who have dreadlocks or beads woven into their hair and wear long flowy skirts. He stole his eyes back to his guitar, fingerpicking.

I leaned back in the chair and took in his room.

It was a beautiful blue-gray color. For being such a hippie, it was immaculate. He had mid-century style furniture and what looked to be like some original artwork on the walls. I scanned the room as he played. Then I glanced at his nightstand, which was only a few feet from me. My mom always said that you could tell what kind of person someone was by what book was on their nightstand. *Shaglandi: The Epic Tale of a Space Prostitute.* Okay, so that one caught me off guard. I expected Kerouac or Jack London or maybe even Carl Hiaasen, but some random book about a space prostitute . . . I guess he is a sci-fi fan.

He saw me looking at the book. "Don't judge a book by its cover, or in this case, its title," he laughed. "It was my first attempt, and a failed one, at writing science fiction, but I keep it there to remind me that at least I did it. I published a book."

"I'm sorry. Did you just say you wrote this book?" I stood up and walked the few steps to the nightstand and picked it up, flipping it over to read the back cover.

"I did. Right after my parents died. I was really struggling. Don't judge me here, but I played D&D as a kid, and that fantasy world had always been a great escape for me. I thought maybe I could escape there again, this time by writing about the characters we used to play. It's a terrible book, but it was super therapeutic to write." He came up to my side, holding his guitar in one hand and looking at the book in my hands.

"Can I borrow it to read?"

"I am not sure it is worth your time, but sure. If you

want to, go for it." He stood even closer to me now. I looked up at him. He had this longing in his eyes, and strangely, it feels safe and comfortable.

He stepped back as I stared into the deep wells of his eyes that seemed to contain eons of wisdom. "Well, probably time to hit the sack. Come by anytime, Mimi. I would love to hear more about your conquest to stop climate offenders, or just about how you are doing."

I tucked the book up under my arm then turned and smiled at him. "Good night, Carter." I closed the door gently behind me.

I found myself struck by how he seemed genuinely interested in what I had to say. I turned back towards the steps, smiling. A sense of calm came over me. At that point, I realized I was ready for finals. *I am ready*. I have tomorrow to finish up my paper and then tests next week and then it was done. Tears fell down my face. Tears of joy, maybe? Or just emotion pouring out of me.

The door behind me opened. It was Carter.

"Hey," he said, surprised to see me still standing there.

"I needed to get some water," he said awkwardly as he stepped towards the stairs, which I was blocking. I wiped my tears quickly, but not quickly enough.

"I have read that tears are just carriers of things we need to get out of our body, so I am not crying, just had some excess cortisol from studying that I needed to get rid of." I gave a small laugh.

Carter's hand softly found my chin and his other hand slowly wiped the tear from the side of my face. "It looks to me like you can use a hug." He embraced me deeply.

I fit perfectly inside his warm cocoon. My face nestled into his chest, his head resting on top of mine. The beating of his heart was steady, a metronome for existence. He gave a tighter squeeze and stepped back with his hands resting on my shoulders.

"See? Don't hugs make everything better?" His hair fell into his eyes as he smiled at me. I could only nod in acknowledgement. The only thing my body wanted to do right now was kiss him. He felt like home. I rose to my tiptoes and leaned in.

Carter stepped back. "Meems, I don't want this to happen when you are so full emotion. I am here for you, that is what friends do."

That was it. Carter just rejected me.

"I know you have someone else, and I am not the kind of guy who is casual about his relationships. I need deep connections. As much as I want to kiss you right now, my heart cannot handle it unless it would lead to something more meaningful. You are a really interesting person who I would like to get to know more, when and if you have the time."

Okay, so maybe not rejected by Carter. Now I am really confused. The words came out of my mouth before I had time to think.

"Yes, Carter, and no. It is a yeah-no situation. I was hanging out with Damian, but that is over. It is really complicated. In this life, it seems we only get a few chances at meaningful connection. The opportunities to meet people are everywhere, but at the same time, it is impossible. Clara has tried to convince me that our collective energy can change things for the positive, or the negative. I don't know why I am saying all of this to you except that I feel like you were put here for a reason. I am not sure what that reason is. But when you are around, that collective energy seems to be positive, and you make me smile. I am a strange bird. I am not sure you really want to get to know me better."

"Good thing for you is that I am quite the hobbyist birder. The stranger the bird, the more interested I am."

"I am not some kinky warbler with a beautiful call, and I have a lot of baggage from recent events that I am working through. And finals are next week."

"Let me take you on a proper date. I can make a picnic and we can go to the arboretum. You have to take a break from studying at some point, and I know you need to eat. Just a few hours. We can see how many bird calls we can hear." He moved in close to me, very close. I felt his energy.

"Not this weekend. I have to focus. How about Tuesday? I have an early final, and nothing on Wednesday, as long as I turn my paper in on time."

"It's a date, then. How about another good luck hug?"

I stepped back into his arms, wishing I could listen to the calming heart metronome all night, but I pushed back quickly and turned and walked down the stairs.

PIVOT AND PLAN

Damian left his apartment to head out to the site of the festival. Planning was in full swing, and they were walking the proposed site today. He was in charge of purchasing and vendors. The event was about two months away and was non-optional. His cooperation would be gained through either the easy way or a beating. Damian pivoted his engagements with the mafia men, letting on that he now saw how prolific the income could be from such an arrangement and that he now saw why his brother was working with these fools.

"I have looked through the estimates and think we have the potential to put at least $500,000, maybe more, through this event unnoticed," Damian said to the man standing in front of him.

"I am glad to see you have come around to see things our way. To see the opportunities, Damian," the man responded.

Damian tried not to appear as uncomfortable as he was. "After looking at the payout, the job is a little hard to turn down. I get to keep the Club and make a little extra money on the side; It works out for everyone."

"If you can pull this off, the sky is the limit." The man turned and walked away, followed by two of the other hitmen. A third man, a tall and smarmy accountant-looking type, stayed behind.

"I am watching you, Miami. Every dollar better be

accounted for. The boss does not like it when people skim off the top. Those people, well, let's just say, they don't stick around long." He pulled back his jacket to reveal the gun on his hip.

"Message received," Damian said, trying not to stare at the gun.

"You use the guys we tell you to use. And if we don't have a guy, it is your job to find one who can play nice. We have two of our guys running the show, you are just there to help them out when we ask," he said.

"Understood."

"And during the event, you see nothing, you hear nothing, you know nothing. If anyone asks, you are only involved because of the Club and you want to bring in more acts. It is a business opportunity."

"Anything else?"

"Don't ask questions and don't talk to anyone. We are watching you. You work for the boss now and he has eyes everywhere. He is a staple in this community. He is a leader. We wouldn't want anything to change people's perception of him being generous and anything but a businessman, got it?"

"Got it."

The tall man turned and walked away, leaving Damian standing there with a tablet that had recorded the entire conversation.

COLLEGE DAYS

I woke up Sunday morning with an epiphany. I am a bit neurotic, a bit obsessive compulsive, and have a tinge of anxiety problem in certain situations. But I am really smart. And all of that studying I was doing, well, I already really knew the stuff. My whole life, I have been able to just remember things. I hear them and they stick. I read them and they stick. The studying wasn't because I didn't know stuff; the studying was an escape and a necessary means to an end. It was what I was supposed to do, or at least what I *thought* I was supposed to do. I may not be the best writer—writing papers is my nemesis—but I remember things. I sat up this morning realizing that I've got this.

The soft pale pink and yellow of the sunrise floated outside of my lead pane window. The leaves on the poplar tree were slightly twisting to their life force. A vireo outside the window made its presence known with incessant singing, but today it was a welcome song, because I am a strange bird.

Today, I was going to finish my final ethics paper and then help cook our final family dinner in the house.

Kiki hadn't been around much since she returned from Brazil. I have been so busy that I have barely had a chance to talk to her. I was excited for our final family dinner, all together, before finals.

The sun gradually moved across the sky as I sat at

my desk, slowly reading and revising my work from the last semester. The current paper was an analysis on government climate mitigation investments in three Kentucky towns and a look at the environmental justice and ethical differences between them. It was not my best work, but it was not bad. Typing out my last sentence, I felt a wave of relief. Closing my laptop, I wanted to do a little dance. My last paper for undergrad was complete.

"I need your help," Clara said, bursting into the room through the bathroom door we shared. Her timing was impeccable, as always.

"I'm all ears, because I just finished my last paper. It is done. At least, I think I'm done. For now." A little shriek of happiness escaped from within me.

"Oh my god. You finished your paper before me. That is a first. Well, I am stuck. I cannot figure out how to end my story. Every way I write it, it ends up sounding like there is no closure."

"Well, you know why I love the Alt+4 on my computer." She looked at me confused. "Because it brings closure."

We both erupted in laughter.

"Your dad would be proud, and thanks, I needed that. I should not have gone out last night. I thought I had it all figured out, and here I am, stuck."

"Well, the best way to get unstuck from something is to completely take your mind off of it. Want to take a walk to The Bean?"

"Hell to the yes. That sounds like a perfect idea."

As we made the familiar walk, she caught me up on

her scholarship and how she and Marcus were doing. And then she stopped mid-stride and looked at me.

"And we saw Damian last night. He was looking real good. And clearly looking for you, too! What happened between you? I thought you had mega-chemistry?"

"We did, but it's complicated, Clara. When everything happened with my dad, I just realized he is not the right guy for me. Don't get me wrong. He is super hot, and, well, a good guy. It is just not the right time. Plus, he is older, has different things going on in life."

"Well, I think he still has feelings for you, and that maybe you need to talk to him, so he isn't pining at the bar for you."

We got to The Bean and ordered a couple of coffees to go. We started making our way back home. Clara nudged my arm. "So, if you and Damian aren't a thing anymore, maybe you should give Carter a chance. I can't stand to watch him stare at you across the room anymore. It's almost awkward. He is such a quirky guy, but I like him. He is super interesting."

"Clara, he is so not my type. Come on. Long hair. Free spirit. For Christ's sake, he wears socks with his sandals. Yes, nice guy, but . . ." My voice trailed off as I thought about telling her. She was my best friend; we never kept anything from each other.

"But what?"

"Soooo, maybe he asked me out on a date, and maybe we are going out this week. But total cone of silence. I do not want everyone in the house to know. And I just want this to be casual, because he is not my type, and I know I *cannot* be his type."

We walked up the stone steps to the front door. Sherp and Kiki were sitting out front on the Adirondack chairs having a beer. We chatted for a few and Clara headed upstairs to finish her paper. I headed to the kitchen to begin dinner prep. It was Nik, Marcel, and I cooking tonight. We planned a big pasta dinner. It was the only thing I really knew how to cook well. Our last family dinner for this semester, and, for some of us, probably our last in the house.

As we all gathered around the table, I realized how much I had come to like this motley crew.

"So, per our tradition, the cooks make the toast. And they," Nik said, pointing at Marcel and I, "told me it was my turn. And, you know me, I don't speak in public, so this will be short. To quote the great Pee-Wee Herman, 'It's been real. It's been fun. It's been real fun.' Now, eat."

"Cheers!" we all said in unison.

I turned to Kiki. "Hey, you alright? I haven't been able to catch up with you since Brazil. Was it totally amazing?"

I saw darkness in her eyes, but the sparkle was still there.

"The first part was a total whirlwind. I had just planned to be on this wonderful jungle excursion, and then the craziest thing happened. . . I was pulled into some investigation by the local authorities to help run some tree forensics. I have no idea how they found me. It took me all night, and I almost overslept and nearly missed the boat."

"Holy crap, that is crazy. But super cool that you got to work some magic while you were there."

"Yeah. Then, the cruise was amazing. We saw so much wildlife. So many exotic trees. I can't put into words the experience, maybe magical. The talk went well. Lots of interest in what we were doing." Kiki's eyes dropped.

"Did something happen?"

"Well, it almost did. I was at the botanical gardens and got mugged walking back. Thank God I know how to protect myself, but what if I didn't? I am still totally spooked by it. It just felt like it was planned. Like it was not random, but I know I am just being crazy."

"Not crazy, at all. You never know. Did you report it to someone?"

"No, I was in a foreign country, who would I report it to? I just got my stuff after it happened and got on the first flight home. I am here and safe now. So happy to be home. I mean it, so happy." She gave her signature Kiki smile, bright white teeth that shone like the sun and made you just want to smile along with her.

"We are glad you are home, too. I am glad my dad made it home safely, too, after all that happened there with him."

"Meems, I haven't seen you since I got back. I heard from Sherp and Nik that your dad ran into some trouble down there. Did he really get kidnapped? Is he okay?"

"He did. And he is. Scariest few days of my life. Remind me not to go to Brazil any time soon. I mean, he got kidnapped and you got mugged."

"What was your dad doing down there?"

"He was doing some research for a non-profit he does some consulting for. He was down some random crazy remote river in the Amazon to connect up with some tribes. Something about extractive industries and their impact on native communities or something like that. I honestly don't know what he was doing there, just that he was with some environmental authorities, and something happened in the village he was visiting. I have learned over the years to not ask too many questions about what he does, as I never seemed to get real answers. I am just glad he is home safe." As I said this, I realized that the whole table was staring at me. Kiki and Clara, especially.

"Wait, Meems, I have a crazy idea here," Kiki said.

Clara jumped in, "Holy, Batman! What if what happened to Kiki and what happened to your dad were related?"

"Don't get ahead of yourself, Clara. What are the chances? They were in totally different parts of Brazil, like thousands of miles away from each other. Statistically, the idea that the mugging and the kidnapping are related, well, it wouldn't even register as remotely related," Nik said.

Kiki leaned forward. "Not the mugging. I struggle to see how that might have been related, too. But in Manaus, that case I mentioned, I was running tree samples to try to find a match for some illegal logs. It was urgent because something was going on out there and they really needed to know species and any identifying info I could give to point to where they might have come from. We were racing against the clock that night. What river did you say your dad was on, Meems?"

"I am not 100 percent sure. They didn't share that information with me. I just know it was in the Amazon somewhere south of Manaus."

Kiki smacked her hands on the table and rose out of her seat. Her eyes wide. "I am almost certain that is the area they said they were looking at for the logging sites. They might actually be related. What if I was helping your dad and didn't even know it? What are the chances?"

Clara chimed in, "Not just crazy, that would make my final project. Well, I don't know, totally flippin' amazing. My whole story is about Kiki and her new tool. To be able to trace her in action to a real-life case like this, it would be huge! I mean if that is okay with you, Kiki?"

"Of course, it is. I was just happy to help. And to put my science into action like that felt really good."

"I will figure out how to ask my dad about this. That all does sound too serendipitous to be unconnected. Also, my mom and dad want to throw a graduation party for us. They want to cater from my favorite Mediterranean place. And Carter, if it is alright with you, they wanted to do it here, in your awesome backyard. All of you would be invited, of course. Just a small gathering of a few other friends, and this group, our little house family. Next Sunday afternoon."

"Of course, Mimi. That sounds wonderful. And maybe we can get to the bottom of the great Brazil mystery when your dad is here." Carter smiled and winked at me.

Monday came and went. A blur of events. The day rushed by but never ended. It was like time was eternal and I could feel the weight of the eternity crushing me just enough to watch the seconds slip by. I had emotions. All of them. Fear, anxiety, excitement, joy. This was it. My last chance to solidify my grades for graduate school. My last few tests as an undergraduate. My life was about to change significantly, and I still was not sure in what way. I floated out of the classroom from my final sociology test. My body was on autopilot, walking the familiar steps, but not directed by any part of my conscious self, so much so that I walked directly into my ethics professor.

"Crap, sorry," I said as I looked up realizing who was the victim of my coffee spill. "Oh no, Dr. Holmes."

Her white shirt was now tie-dyed with chocolate-colored coffee. She flashed a smile at me. "Well, if it isn't Mimi Martin."

"Dr. Holmes, I am so, so sorry. I wasn't paying attention. Crap! I don't even have any napkins or anything."

"Mimi, it is fine. I have another shirt back in my office. This isn't the first time this shirt has met coffee up close and personal. This is good, though. I was wondering if I would see you again this semester. I have been thinking a lot about you lately. Have you made any breakthroughs?"

I stared at her. Clearly, I have not.

"Well, I did finally figure out that I have an excellent memory, I like things in action, and I am a terrible writer. The irony. My mother is a famous author, and here I can barely crank out an undergraduate paper. Other than that, still trying to figure out who I want to be when I grow up."

"Well, I look forward to reading your paper. We all are constantly growing as writers. No one wakes up a good writer. It takes practice, editing, and more editing. Writing consumes more of my life than I would like it to, but, then again, publishing is part of my job. Keep in touch, Mimi. And congratulations. You should be really proud of how you showed up here at this school." She grabbed the front of her shirt, trying to keep the coffee from soaking through then she smiled at me and stuck out her other hand. I took it and looked her right in the eye.

"Thank you, Professor. I really enjoyed your class and your approach to thinking. And your advice."

As I walked away, I was now in my body, in control of my mind, and felt like I could take on the world. I decided I needed to pay Damian a visit. I was going to stop by the Club this afternoon. They aren't open on Mondays, but he was usually there taking stock of stuff or working on repairs. I wanted to see how he was doing and let him know that I would likely be leaving town soon. I didn't feel like we left things in a good spot. Not that we were ever anything to begin with, but Clara talking about seeing him the other night really jarred something in me.

I stopped by The Bean to grab another coffee given mine was now mostly on Professor Holmes' shirt and then headed to the library to put the finishing touches on my ethics paper. I loved this building; It was shaped like an octagon with massive windows on all sides. It was six stories and filled with gorgeous limestone laden with fossils. And there were so many books as well as so many places to tuck yourself away to study and read. It was my safe place, my happy place.

I gifted myself the time to take one last long walk through the building, running my fingers along the rows of books. I spied two young lovers kissing the next row over and quickly moved on. I will miss this place. After a few hours, I knew I could do no more. The paper was officially done. I didn't have to turn it in until tomorrow morning, but I was never one to wait until the last minute.

I wrote an email to Professor Holmes apologizing again for the coffee mishap and thanking her. Then I attached my essay and clicked send. Done!

I packed up my bag and walked across the green open space that surrounded the building, heading north up the main street to the Club to see if I can find Damian.

The lights looked like they were off and the door was locked. But his motorcycle was out front, so I knocked just in case he was in there.

A look of surprise crossed his face when he saw me at the door. And then the look turned to happiness and then concern.

"Meems, hey. So good to see you. Wow, you look

really great. Um, what are you doing here?" He stepped in to give me a kiss on the cheek. This was one cultural difference I never reciprocated. The whole kissing as a greeting instead of handshake, hug or bow was odd to me.

"Hi, Damian. I just wanted to stop by and say hi. Things seemed really weird the way we left it at the coffee shop. I was worried about you. Plus, I wanted to say goodbye."

"Goodbye? Here, come on in." I stepped into the dark bar. He closed the door behind me.

We walked over to the back office behind the bar which was lit by a few lamps. He motioned for me to sit on the couch. He plopped down next to me. I stared at him for a bit. The chemistry was still there, but felt different. His energy was different.

"Are you going somewhere?"

"Well, maybe. I graduate this weekend."

His eyes lit up. "That is right! How could I have forgotten? Congratulations."

"Thanks. I just don't know what is coming next for me. Things have been so crazy. I am trying to decide what to do next, but I don't think my next step is to stay here in town. My lease is up in a few weeks, and with Clara leaving to go to New York, I think it is time I moved on, too."

"Wow. I am happy for you. I—"

The front door slammed open. Heavy footsteps stomped across the floor.

"Barman, you fucked with the wrong person."

A very large man covered in tattoos stood in the

doorway. He looked at me and then at Damian. Damian had already jumped to his feet and ran to the door, trying to intercept him.

"Who is that?"

"That is no one, just some girl who likes to get a little Latin action, if you know what I mean."

I scoffed loudly. "Really? Is that what—" I stopped midsentence as I saw him glare at me.

"Well, she can't be in here." The big man grabbed me by the arm forcefully.

Before I knew it, I was wrapped up in his bicep, my feet barely touching the floor. He started to drag me away. Then my training kicked in. I spun out from under his arm and drove a hard side kick into his kidney. He winced and lunged at me. The palm of my hand met his nose, driving it upward into his forehead. I knew that would stun him for at least enough time for me to get away. Out of the corner of my eye, I saw Damian staring down the barrel of a gun. I turned and ran out the front door then heard the sound of a loud pop. I kept running as fast as I could, straight back towards campus, where I hoped I would get lost in the crowd of students.

As soon as I got to the closest building, the chemistry building, I slipped inside, straight to the bathroom, and slammed the stall door shut behind me. I climbed up onto the toilet and dialed 911. About five minutes had passed by this point.

"There has been a robbery at the Club on main. I think I heard a gunshot. Please send someone quickly." I hung up. The operator tried to ask me all of these questions, but I just wanted to call Anne. I dialed her number.

She answered on the second ring. "Anne, something has happened to a friend of mine, the one you were asking me about a few weeks ago. Someone needs to get to the Club and make sure he is not injured."

"Whoa, slow down, Mimi. Tell me what happened."

"I will, after you get someone on this. Call me back when you do."

My other line was ringing. It was 911 calling me back. I answered.

"Are you in danger?"

"No, I am fine and safe. I will talk to the police, but for now, just get someone to the Club."

I hung up on her again.

I was fully hyperventilating. I took deep breaths to try and calm my breathing. My phone vibrated. It was as if time stopped. I was so confused about what happened. I felt the vibration again. I looked down and saw it was Anne. I answered.

"Anne, I don't know what is going on, but I am scared."

"Where are you now, Mimi? Are you safe?"

"I am hiding in the bathroom of the chemistry building on campus. I think I am safe. But I am sure he is not safe. I know you asked a lot of questions about him and you know who he is, so I called you. I did not know who else to call, besides 911. I called them first. I think he's in some major trouble."

"Stay where you are, Mimi. I am coming to pick you up. I will send someone down to the Club to check out the situation."

It seemed like hours passed in the twenty minutes it took her to arrive. During that time, I ran through a lot of emotions. Fear was the undercurrent, but also confusion and anger. I collected myself. I was still squatting on the toilet and didn't even realize that my legs had fallen asleep. I almost fell over when I stood up. I heard Anne calling my name and slowly opened the stall door. I told her what had happened as we walked to the car.

"Describe for me whatever you can remember about the man who grabbed you," Anne said, starting at me with her intense brown eyes. She was in plain clothes. I must have caught her off duty.

I told her everything I could remember. How big he was. The short man pointing the gun at Damian.

"Did they say anything about why they were there?"

"He just said something like, bar guy, you messed . . . No, he said, 'you fucked with the wrong people and then everything happened." Just relaying these words to Anne brought back a flush of emotion.

"Something is not right there, Anne. Damian is a good guy. This whole thing is just not right."

CATCH UP

Niles knew he was nearing the end of this assignment, so he checked in with the attaché, Clyde, in Vietnam a few times over the last month with follow-up questions about the country's new import and trading laws. He was also trying to find information on what had happened to Thí, the non-profit director who had lost her life the day he met her. Clyde had been only a little helpful, less than responsive, and, overall, a bit glib. Niles wasn't exactly sure how he had landed this semi-coveted post until he opened his email that afternoon. Clyde had sent him an encrypted email earlier with some documents and a new assessment report that looked promising.

Niles combed through the second report when he noticed there were random words interspersed in various paragraphs. They were all related to the topic, but just out of place. *Poor editing,* Niles thought to himself. Then he realized that the placement was consistent throughout the document. He started to pull out the words, writing them on his steno pad.

Internal. Threat. Corrupt. Senior. Officer. Deputy. Shipping. China.

If what he was reading was correct, the Deputy Ambassador may be to blame. Niles hadn't met him during his time in Vietnam. For Clyde to send the message like this, it must mean that he was afraid. Niles knew better than to call him and discuss this revelation, but he could call him to discuss the assessment. Niles rang Clyde, the phone rang three times.

"Spees here."

"Clyde, hi, it's Niles Martin. I wanted to follow up with you on that assessment you sent over. I appreciate you sharing the details of the most recent seizure. It looks like things are still going strong over there, despite all of the new laws and regulations."

"Things are happening, even if it might not seem like it to the outside. We are talking about millions of dollars of lost revenue to legitimate sources. The US and Vietnam take this very seriously. The task force just needs a refresh. Some new faces and new ideas." He sat back in his chair to appear relaxed even though his heart was racing and his palms sweating.

"Oh, right, of course. Like people from non-profits. Speaking of which, any word about the non-profit I visited?"

"Mr. Martin, you of all people should understand how this works. There is nothing I can speak about today. Word on the street is that this was a local dispute," Clyde spoke in a louder than normal voice in case anyone was listening in on the other side of the paper-thin wall that separated his office from one of the smaller conference rooms on one side and the political attaché's office on the other.

"Right. Well, I just wanted to let you know that we are still working to bring justice to impacted communities and environments and appreciate your help to better understand the issues."

"Of course, it is my job, Niles."

Niles asked a few questions about numbers in the report. Clyde regurgitated information like it was on speed dial in his brain.

"Clyde, thanks for the report. There was a lot of really great information in there. I have read it closely. I will be sharing it with my colleagues, and we will see what we can do to help." Niles hung up the phone.

The cagey answers and tone of Clyde's voice told him everything he needed to know. Maybe this young foreign service agent was more adept than Niles gave him credit for. After all, it takes serious guts to do what he did. And then to answer the phone and play it off like that.

He picked up the phone to call David and noticed that he had two missed texts from him.

Everything is okay. Need you to call me. ASAP.

We need to talk. Call ASAP.

Niles dialed David and it went to voicemail.

"David, it's Niles. I've got something. We need to talk. Sounds like you need to talk, too. I'm coming to Lexington now. Let me know where and when to find you." He hung up the phone, collected his files and laptop, and went to talk to Hae.

He found her in the kitchen, pouring tea. As he walked in, she turned to smile at him. He stopped and smiled back. This is what he wanted. He was ready to have these moments every day, to watch her so peacefully stand over her tea mug to watch the leaves bloom. To be home. To be with her, his wife of 25 years.

She understood why he did what he did. She had

always been by his side. She knew his passion for the work, his call to service, his commitment to country, and his commitment to family. And he knew her love of writing. How she would never stop, nor should she. She was brilliant. It was her lifeline. Her characters came to life in a way that writers dream of. She knew them each and spoke about them as if they were old friends. He would never ask her to give them up. They would share dinner with them every day until the end of their time, and he supported that. He was the one who needed to change, and he was glad his last case was just about finished. At least, he hoped it was.

"Another trip to Lexington, my love. I think we are close to wrapping this one up, so hopefully it is my last."

She looked over at him, the steam from her mug rising in front of her. She gave him a firm look. "Niles, please be safe."

TRAINING PAYS OFF

The rest of the day was a blur. I was terrified about what happened. I now understood why Kiki just wanted to get the hell out of Brazil and not talk about her assault. I wanted to curl into a tiny ball and not be seen until I knew if Damian was okay. There were no words to describe what happened. It was all so fast, so unexpected. I was back at the house and, once again, there was an unmarked car sitting outside keeping watch. I was lying in my bed, the sun was on the other side of the house now, my room was finally cooling off.

The door to my room flew open and Clara bounced in.

"Meems, I crushed that—oh, wait, Meems, are you okay?" She stopped mid-stride seeing me there face down with a pillow over my head.

I could be exactly who I was with Clara. I did not have to pretend to be tough. Half the time, it was like she could sense what was going on with me before I could. She came and sat on the bed and slowly lifted the pillow off my head.

I rolled over and pushed myself up to sit, wrapped my arms around my legs, and tucked my knees into my chest.

"Clara, today. Today. Um, today . . ." She sat there staring at me, nodding, then reached over to give me a hug. As I rested my head on her shoulder, I let it all out.

"Clara, Damian and I were over. I don't know why I went there but I did. I felt empowered to just say goodbye again and see him again. I don't know what came over me. I went to see him and then these guys came in, and they grabbed me, and they had a gun, and then I ran, and I heard shots. And I don't know if Damian is okay or not."

"Oh. Wow. Holy shit, Mimi."

"I think he is dead. Or at least really hurt because I went there. It is all my fault."

"Whoa, wait, what? How?"

"If I hadn't been there, maybe they would have just robbed the place or done whatever, not shot him." My veins felt ice cold. I realized I was not crying. I was now angry. Angry at the men. Angry at whatever Damian was up to that put him in that situation.

"Meems, this is not your fault. Whatever brought those guys there, they were there not knowing you were. They were going to do whatever they were going to do. Did they hurt you?"

"A little bruised, but not hurt. Clara, I broke that dude's nose and it felt so good. Thank God for all of those years at the dojo. I never thought I would have had to use those skills in real life."

"Hell yes, that's my girl! What else happened? Did you talk to the police?"

"Yes, of course I talked to the police, but I still don't know what Damian's status is or what happened to him. It feels like it is all this horribly difficult 1,000-piece puzzle where half the damn thing is all the same color and I am trying to piece it together, but the end pieces are not in the box, so I don't even know where to start."

"Okay, well, you are safe here now. I am sure you will hear something soon. No one will hurt you here. You are in a house full of peace-loving hippie ninjas who would do anything to protect you. The biggest oxymoron one could imagine, but hey, it is reality. Do you need anything?"

"No, just to not freak out while I wait to hear about Damian. So, you came in here to tell me something. Please tell me something. Tell me anything to get me to stop thinking about this." I lifted my head off of her shoulder and forced a smile.

"Oh, it is nothing. I just aced my test and finished my paper. I am almost totally done. Only one more test to go. It just feels surreal to almost be done."

"Crushing it, Clara, like you always do. I don't know how you stay so damn positive all the time. No matter what is happening, you just have this perspective."

"Meems, it is pretty easy, honestly. Everything we do, everyone we meet, there is a story there. Stories are meant to teach us. So, I just look at every moment, good and bad, as an opportunity to learn and grow. And when I do that, I can usually find something good. Plus, for me at least, it is way easier to keep things light and sunshiny. My soul cup gets filled that way, and lord knows it needs a lot of filling," she laughed, and I laughed with her.

"Hey, so, truth be told, I had also come in here to ask about your big date with Carter tomorrow. Are you still going to do that with all that has happened?"

My eyes widened. I hadn't even thought about that

after everything. "Ugh. No, I can't in good conscience go on a date with Carter after what is going on with Damian. That's not fair to him. And not fair to me. Last thing I need is another set of my puzzle pieces to go missing in the confusion. I have to figure out how to tell him."

"Meems, I know you and Mr. 10 O'clock had some crazy chemistry, but I think Carter is a really good guy. I think you should at least give him a chance. He is so not your type, but maybe that is just what you need. Just don't break his heart."

LEXINGTON

David listened to Niles' voicemail inside the Club. About two hours had passed since the incident yet David still hadn't had a chance to connect with Niles. The police had come and gone, and he and Anne were wrapping up their survey of the damage. He had only been back from Brazil for a few days. He had traveled to a few more locations collecting beta, but still nothing concrete. He knew they were getting closer. All of the indicators pointed to a well-organized and well-funded syndicate, possibly even more than one. He was back in Kentucky now to wrap up their work here. He was starting to seriously consider retirement after this operation was over.

"Anne," David whispered under his breath, in case the place was bugged. He showed her the voicemail transcription from Niles. She nodded.

"We are almost done here, anyhow. We can go back. Just one more walk through of the office."

The office was a mess. The couch cushions were unzipped and unstuffed. The bottom of the couch was cut out. Files scattered across the office. Whoever was here was looking for something specific.

Anne looked down and saw a small air vent next to the desk, barely visible. If she were looking to hide something, where would she put it? They had already moved the ceiling tiles, nothing. She grabbed her multitool and unscrewed the sides of the air vent and the cover dropped. She grabbed her flashlight and shone it around, nothing visible. She reached her hand inside and felt around, nothing on the bottom or sides. Then she felt a small bump on the top. Peeling back the tape, she pried loose a flash drive. Bingo.

She looked over her shoulder at David, showing him the flash drive. "Nothing here," she said. "Let's go."

Back at the office house, David and Anne had the thumb drive scanned by the IT team and were awaiting information on what files it contained. Niles showed up about 30 minutes later.

"Niles, it is good you are here. A lot has happened today," said Anne with her brown eyes set intently on the laptop in front of her. The files were starting to come in from IT.

"Same, I may have found something, a lead for Vietnam."

The three of them sat around the table.

"Listen, Niles, I tried to get to call and text. You need to hear about what happened today. Before I start, Mimi is okay," David said, sitting across from Niles. Anne's eyes lifted away from the laptop.

"Mimi? What does she have to do with any of this?"

"Niles, Mimi was involved in an incident today. She got very lucky. Her friend, Damian, was not so lucky. He is in the ICU right now. The doctor's think he will pull through, but it is still hit or miss. We have two agents there on standby to make sure no one comes to pay him an unexpected visit."

"Tell me what happened," he demanded.

"Niles, Mimi called me around 2 p.m. today. She had gone to talk to Damian when some suspects entered the Club and threatened them. She was able to use her self-defense skills to get away. The video footage of her working the guy over is impressive. She ran to the closest place she felt safe and called me. She has a few minor bruises on her arm. Shaken up, confused, but okay."

"We spent the last hour at the scene cleaning up after the police trying to figure out what the guys were after. The police wanted to call it a robbery because the register was emptied. But with the file and office destruction, it was clear they were looking for something else. And we think we found something else," David said as he stood up and walked around to see what Anne was looking at on the screen.

"Niles, I know how hard this is for you, but you cannot let Mimi know you know. It will put you too close to this case, and we can't have that when we are this close. We have someone watching her," Anne said, pausing for a moment so Niles could absorb what he had just heard. "I know how hard this must be for you."

"I understand. But I am going to call her now. Just to check-in. A friendly dad call."

Niles looked at his phone and dialed. She answered on the first ring.

"Hey, honey, do you want a good pickle joke that you are really going to relish?" He tried to keep it light to gauge her mood.

"Dad, hey. Okay, hit me with it." He could tell that she was smiling.

"Well, you know I once tripped over a pickle and fell. I am over it now, but it was a really big dill at the time."

"Bu-duh-dun-sh."

"Just checking in on you, honey. I know it is your last finals week. Your mom and I cannot wait to come up and see you this weekend." He waited. Would she tell him what happened today?

"School is fine, I am fine. Everything is fine. Today was just a really crazy day. I am okay, but still processing all of it. I don't even know how to put it in words yet. I was almost attacked, or maybe I *was* attacked, I don't know. I was able to get away. I'm not hurt, nobody did anything bad to me, so don't go to the freakout zone. But my friend Damian, the one you met, these bad guys, I don't know if they were robbers, or if he owed somebody money, or what he was into, but I think he might have been shot. I called that really nice agent who helped Mom and I when you were kidnapped and asked her to see if he was okay. It has been several hours and nothing. So, I am kind of freaking out about that. But I am safe in my room here and home. Clara is here. Everyone is here."

"As your father, I am pretty freaked out about you being attacked. I hope you told the police everything you knew. And lay low for a few days, okay? I am going to come up to see you tomorrow."

"No, Dad. Really, I am fine, and I will be a lot better once I know if Damian is alright. I have tomorrow off, no exams. So, I am just going to stay home and hunker down."

"I'm just a phone call away if you need anything at

all."

"Okay."

"Mimi? Still there?"

"Yeah, Dad, um, I have a question. . . So, my friend, Kiki, was in Brazil at the same time as you. She helped on forensics to trace some wood for an urgent situation that just so happened to be the same days you were being held hostage. Then she was attacked in the arboretum when she was at the conference. I can't help but wonder if she wasn't working on helping to find you."

Niles sat there trying to conjure up the right words to tell his daughter. He remained silent.

"I feel like all of this stuff is connected somehow."

"What would make you think that?"

"Just a hunch, Dad. And this whole thing with Clara and Marcel. And the thing with Damian and I today. The thing with Kiki. It is almost unbelievable, right? I mean, statistically, what are the chances of all of these awful things happening to people I know and love so close together like this? These things don't happen to people. They are things that happen in movies or in books, not in real life. Why are all of these things happening around me?"

"Mimi, this sounds like a bigger conversation that we should have in person. I am just glad you are okay, honey. I cannot even imagine how you are feeling right now. I can be there in the morning for coffee. If you have the day off, we can talk through everything, okay?"

"Okay, Dad, see you in the morning."

JACKPOT

Anne and David both looked at the screen, scrolling the contents of the thumb drive for anything relevant. It was account information for the books for the Club, which included transaction registers and a stream of account numbers for various banks.

"Jackpot!" Anne exclaimed. "I am going to send this off to forensics and the analytics team to see if we can get some traces on these accounts and see who holds them or where they go."

Niles came back into the room. He had just hung up after talking with Mimi.

"I need to tell you all what I found. I received insider communication from Vietnam implicating a senior US official within the Embassy as being corrupt and working with China."

Anne and David stared at him, but David spoke, "What evidence do you have of this?"

"Right now, it is just insider information. I am working on getting something more concrete. For now, I think you all need to have someone keep a close eye on what is happening inside that Embassy." Niles stood to leave.

"This is starting to make sense. In Brazil, one of the agents shook my hand and gave me a coin. It was a State Department challenge coin for the Guangzhou consulate. I held onto it not knowing what to make of it at the time. I figured it was related to the case but could not figure out how. I am not sure it is some coincidence that both your source and mine tie to China." David stroked his salt and pepper beard.

"Let's not jump to any conclusions here. We have two very different sources of, at best, unvetted intelligence. They may be connected; they may not be. Let's keep working all the angles here for our local case and we will get this info on the bigger case up to the DC team." Anne opened the garage door to leave.

"Anne," Niles called after her. "I think Mimi trusts you. Please keep her informed about what's going on with her friend so she doesn't worry too much."

She smiled back at him, "Of course Niles."

Niles waited several minutes before retracing his steps back to the car. Every part of him wanted to see his daughter. He drove by her house just to make sure someone was stationed out front. When he confirmed she was safe, he headed for the Hilton downtown, hoping they had a room available for the night.

Anne picked up the phone to call Mimi, she answered on the first ring.

"Anne, hi. How is Damian?"

"Mimi, the sound you heard was a gun going off, and Damian did take a bullet, but he is going to make it. He can't have visitors right now, but the doctor's say maybe in a few days. If anything changes, I will let you know. He was lucky. You were lucky. Lay low, kiddo, and keep your head on a swivel for now. We don't have any reason to think you are in danger, but just keep your eyes open, okay?"

"Okay, and thanks, Anne."

The line went dead.

MEANWHILE, BACK AT THE RANCH

The car whipped into the driveway as if they were being chased by a herd of water buffalo, or maybe the police. Yet, no one was chasing them except their own conscience. Pulling into the large garage next to the barn, the car was now out of sight. The two men stood in the garage corroborating their stories before casually walking down the wood-chipped path to the barn at the back of the estate. The police sirens in the distance forced them to get out of the Club before they could erase the camera footage, but at least they were able to kill the cameras before they left, and they had taken out the one that watches the parking lot weeks ago.

This was the secondary barn. It was for the lead ponies. The thoroughbreds had a virtual mansion up at the front of the property near the real mansion. It was quiet at the back end of the estate, surrounded by rolling hills and dense deciduous forest. Stand still too long and you would become part of it. In fact, many bodies have fertilized those woods over the last decade.

"Why did you shoot him, you dipshit?!" said the tattooed one. "I could kill you myself for complicating things like this. But I won't have to, because he will probably kill us both thanks to your fucking idiotic trigger finger. If he is dead, this shit is going to blow up and you know it."

The short and stocky one remained silent, knowing

his fate. They stood at the back of the barn waiting silently for their boss to arrive. This was the routine. He would make his rounds to see his horses, and he would stop in here to provide whatever job was next or collect any information he had asked for. The door to the barn slid open.

The personal assistant walked forward. All six foot six and 300 pounds of him. He nodded and weapons were handed over. He paused. He then began to pat them down. This was out of the ordinary. Obviously, the trust level was low around here. When he finished his search, he turned and walked back to the barn door, his frame so large he almost blocked the low sun.

Carrying his walking stick, which was unnecessary as he was uninjured—he only carried it to make a point—the boss slowly walked towards the two men. When he was within ten steps, he made a tsk-tsk sound, like he was scolding two young children for playing with something inappropriate.

"Boss, listen, we can explain."

"Oh, I am in need of some entertainment today, it was quite the boring day up until a few hours ago. Please, regale me with your stories of heroism and triumph."

"It was a damn girl. She wasn't supposed to be there. We were just going to deliver the message and a few punches to drive it home. You know, stop poking your head where it isn't wanted. But then there was someone there. And she just triggered this whole fucking thing and then next thing you know, she is gone, and I had no choice but to shoot him before he shot me," the short one paused and then continued when the boss stood looking only amused at him.

"I mean, he was a piece of shit, anyways. He didn't know what he was doing and was only going to cause problems for us. He is not company material. It is better that he is out of the picture."

"Out. Of. The. Picture?" His checkered traditional ghutra headdress cast a wide shadow as he leaned forward, continuing. "The only ones who are about to be out of the picture are you two incapable idiots. *He* is not out of the picture. *He* is in the hospital and will likely end up being just fine. Do you know what that means? My job just got a whole lot more complicated. Because now he is surrounded by police. And what do you think he is telling them about who shot him and why? So, if you want to remain with the walking, you will go and take care of him so that he cannot talk when he wakes up from his little nap."

"Boss, we won't let you down."

"I have very important guests coming in this weekend to visit. I do not need this nuisance of a situation getting in the way of my business. I do not want to hear from you or see you until this is done. By the time I wake up tomorrow, this will all be handled, or you will be handled." He pointed his walking stick at the two of them, revealing a gold cap on the bottom, which likely sheathed a blade.

The white robe, which normally was silent when he moved, seemed to echo ripples of energy through the barn as he turned, whipping it with him. He climbed into his golf cart and proceeded up the path towards the main barn.

The pro wrestler-looking bodyguard remained watching them, holding their weapons.

"I look forward to seeing you tomorrow," he said as he dropped their weapons on the ground, turned and walked out the barn door, climbed on his dirt bike, and followed his boss up the path.

IT'S THE MORNING, SUNSHINE

It was a night where I could not sleep at all, and yet, I slept deeply. I oscillated between exhaustion and anxiety. When those first rays of sun came through my window, I knew I had to get up. I had to move on. I had to keep being. I was stronger than this. And I was incredibly curious what my dad would share with me today. I showered, dressed, and went to the kitchen for some coffee. Carter was at the sink, his back to me. I saw the picnic basket laid out. *Crap*. I had totally forgotten about our date today. I wanted to turn and run but it was too late, not an option.

"Well, good morning. Look who is up with the sun today."

"Hey, Carter. Yeah. Um, so, I am not sure our hanging out is going to be able to happen today." The look on his face, once content, was now apprehensive.

"My dad is going to be here soon, and he and I really need to talk. It's a long story, but a lot happened yesterday. I am really sorry," I said as I looked into his eyes, I realized how confused I was. He was not angry. He was concerned.

"Meems, it's okay. Are you okay? You look like you have seen a ghost." He walked around the kitchen island to stand next to me.

"I pretty much did. Or I am living with one right now. One minute, life is all fluffy puppies and rainbows, you know, picture perfect. Exactly how it is supposed to be. And now, I don't know which way is up."

"You have found life. Real life. It sounds. Each moment of darkness will eventually be greeted by the light. The walk through the darkness is never easy. Lots of people get lost in it. Have you ever read any Ursula Le Guin?"

I just stared at him. I nodded, wondering where he was going with this.

"She has this great quote: 'People who deny the existence of dragons are often eaten by dragons. From within.' Don't let your inner dragons eat away at you, Meems. Acknowledge them. Get to know them. And, eventually, they will become your teachers. And then you will know which way is up. Am I getting a raincheck for the date?"

"Raincheck, yes. Thanks for understanding, Carter." I finished pouring my coffee into my travel mug, gave him a genuine, but weak, smile and turned towards the door.

My dad texted that he was out front.

I opened the passenger door to his Audi Q3. I loved this car. "Morning, Dad."

"Hey, sunshine."

He looked like his usual purposefully messy hair, shirt slightly untucked, professor-style disheveled self, which was no indication as to how he was feeling. But his eyes looked tired and sad. I knew what I had to do.

"Oh no, Dad, the sun!"

He smiled at me. "The sun! What is it?"

"Well, it is a large fiery ball at the center of our solar

system, but that is not important right now." Even saying it made everything felt right. When I was in high school, we used to watch slapstick movies, and *Airplane* and *Airplane II* were on repeat as favorites. We could recite them with ease.

Just having this moment of normal felt so good. Maybe Carter was right. There was a way to tame the dragons, to see the light even in the darkness.

"Where to?"

"Let's go to Waffle House. I haven't been there in ages, and I would love some smothered and covered."

"Roger that, kiddo."

"So, Dad, I heard from Anne last night that they thought Damian was going to be okay. I can't help but wonder if that trouble he was in, or whatever it was that he wanted to meet with you about that day at the Mediterranean restaurant . . .well, I wonder if that is related to what happened yesterday."

My dad kept his eyes on the road. No response.

"I mean, it cannot be just a coincidence that a few weeks ago he is scared, beat up, needs help and then yesterday he gets shot. I don't want to know what he was involved with, Dad. I just want to know that whoever came after him are not going to come after me now because I was there. I cannot help but feel that everyone I love right now is in danger, and I don't know why or how."

"I hope this never changes, Mimi. You talking to me like this, trusting me. I can imagine how scared you are, how upset you are, how many questions you have."

I hold back the fountain of tears brimming in my eyes

and am thankful that I didn't put on any mascara this morning.

"Mimi, sometimes questions are better left unanswered. When you know something, you cannot unknow it. Just like when you see something terrible, you cannot unsee it. Knowing things can be unsafe. I am not sure you want to or should know or can know what is really going on. I think it will be much safer for you to not know and to not ask questions."

"What? Dad. Really? You expect me to just sit here and not ask questions? I mean, Damian and I only barely knew each other, went on a few dates, hung out a few times. But I liked him, and he liked me and he was my friend."

I switched from fear and sadness to anger. I felt it rise in my body and realize that I am much more in control of my emotions right now than I expected. They are not a volcano erupting. They feel more like an iceberg. I am able to only show the tip of them and just feel all the rest inside, unseen.

"Dad, I am going to ask questions and if you know the answers, I would really like to hear them. I am really sorry that I got you involved in this. I know you have been busy with your volunteer work and this contract case, and then I dump this Damian thing on you and now, because of me, I am guessing somehow you have to deal with this."

"Listen sunshine, there are things I just cannot tell you. I am sorry. I can't. For your safety and for other reasons. But there are other things I can share. So, how about you ask me one question at a time? If I can answer it, I will. If I can't, you are going to have to accept that and understand it is for your own good. Deal?"

I brooded in my seat, realizing I had no option but to accept this deal. I knew my dad would only want the best for me and he had reasons for whatever he was doing.

We pulled into the Waffle House parking lot and he turned off the car.

My father shifted in his seat to look at me. "We are not doing this in a public place. Breakfast will have to wait until after we get through my interrogation."

ATTEMPT

It was about 2 a.m. when they arrived at the quiet hospital. An ambulance sat in the emergency pull-through with its lights on. The parking lot was dark with no pedestrian traffic. The short one pulled into a spot at the edge of the lot. The tall one got out of the passenger seat and walked towards the entrance.

This would be an easy job. A quick walk upstairs, a quicker injection into his thigh, then a sprint back down to the car. An easy job. He had been here at this hospital a few times before. Various reasons. The birth of his child, whom he hadn't seen since, was one. That time, he was shot by a "stray bullet." He knew the back stairwells and the various ways to remain unseen.

He had not anticipated the number of people on the ICU floor. It was like a mass casualty had happened. Nurses walking swiftly in every direction. The incessant sound of beeping filled the room. Always beeping, the lifelines, the drip bags, the temperatures, the blood pressure. Everything had an electronic sound, and it seemed they were all alerting at once. But, this time, the noise was welcome.

A group gathered at the check-in desk discussing charts. That desk was directly across from the target's room. There was no way to get in there unnoticed. He walked down to the break area and poured himself coffee and then headed back down the stairs, making a lap through the cafeteria and back up the opposite stairwell, hoping the desk gaggle had broken up to go do their jobs.

Round two of the walkthrough, he saw a man sitting in a chair just outside the door of the room. He had missed this before with the chart carts being parked in front of the room. This was going to require more skill than he was prepared for. He walked back down the hall and followed a nurse into the bathroom. Four minutes later, he came out in scrubs that were significantly tighter than he was comfortable wearing. Between the mask and the scrubs, most of his tattoos were covered.

He walked with authority to enter the room, pushing the chart cart in front of him. The officer at the door stopped him.

"ID?"

The large man pulled the lanyard from around his neck to show the guard.

The guard looked at the picture, looked at the man, looked again at the picture.

"You aren't the same guy who was just here. Where did Jeremy go?"

"Oh, uh, break. I am just taking this one for him."

The guard looked the big man up and down.

"Hold on a minute, I have to register your badge number." He picked up his phone and a clipboard, asked to see the ID again, typed something in, and then wrote it down.

"Let's go, then." The guard opened the door to usher him in.

Way easier than the big man could have imagined. Except then the guard followed him in.

They walked to the side of the bed. The big man stared at the machine and then looked down at Damian.

"I got this. You are good to go."

"That isn't how this works. I am supposed to watch everything that goes on here. I guess this guy is somebody important."

Suddenly, the door slammed open and two plainclothes men and one woman entered with guns drawn. Within an instant, the big man was on the ground. He hadn't even had time to reach for his gun.

Meanwhile, Damian lay unconscious, machines beeping all around him.

Q&A

Staring intently at my father, I carefully considered my options. What did I need to know?

"What were you doing in Brazil?"

"I was conducting research into illegal timber trades in the Amazon."

"Why would that end up getting you kidnapped?"

"I was in the wrong place at the wrong time."

"Did Kiki have something to do with your rescue?"

"As I understand it, Kiki did conduct forensic analysis on a seized wood shipment and her incredible skill and analysis did help determine the general area where I was being held."

"How did they know about Kiki, that she was there? How did they find her?"

"You had talked about Kiki's research, and it is well known in the scientific community."

"You didn't answer the whole question, Dad."

"I answered what I could."

I paused, considering where to go next. "What was Damian into that he needed your counsel on?"

"You know I cannot share that with you, honey."

"Was Damian in danger when he reached out to you?"

"Danger was a high possibility."

"Were you helping him?"

"I was doing what I could from the position that I am

in."

I paused again. My heart pounded in my ears. I wanted so many more answers than I was getting. I knew this session with my dad was not going to yield what I needed. I needed to talk to Anne.

"Are you afraid for me right now?"

He shifted in his seat and dropped his gaze. That was an answer enough.

"Are you really working for a non-profit?"

No response.

"Are you in danger, too? Is Mom?"

"Not at this time. Not that I am aware of."

Then it hit me. "Did you know about Damian being shot before I told you last night?"

No response.

"Dad, I am scared. I don't know what is going on, but I want it to stop. I want my normal life back. I don't want my roommates and friends to be involved. I just want this all to stop."

He reached across and grabbed my hand.

"It will all be over soon, honey. And I promise you, everything that can be done is being done to ensure you and your friends are not in danger. Let's go get some terribly fried and greasy food."

Dad's phone pinged as they walked into the smell of deep-fried and buttery goodness. It was Anne. It would have to wait until after breakfast.

"Meems, do you know what dogs eat for breakfast?"

The eye roll was subconscious and automatic. He didn't wait for me to respond.

"Woofles."

Interrogation of an Imbecile

"I told you, I don't know."

"Really, you don't know who hired you to try to give an excessive dose of propofol to the patient?" Anne questioned the man. The local police had already had their go-round with him and would likely have another after this.

"I don't know. It was one of those anonymous jobs. Hired through word of mouth. The guy must have just pissed some people off. I don't know."

"Well, you know what I know. I know that you are being charged with attempted homicide. There is really no way out of a guilty conviction for you. Do you know what that means? A minimum of 16 years in prison, but more than likely, looking at you, it would be the rest of your days in a dark, damp, concrete block with terrible food and even more terrible conditions."

"I have been there before. I know what the place looks like. It doesn't change what I am telling you. I don't know who the job was for."

"Well, that is a real shame, because if you did, we might have been able to work out some kind of deal for you." Anne stood, pushed back from the table then turned and walked out of the room.

From behind the one-way glass window, Anne and the local officer chatted. She made a couple of calls to DC and then talked to the sheriff.

"I thought you had him," said the man in blue.

"I think I do," responded Anne. "He is as close to ratting as anyone. I can see it in his eyes. He just needs a few minutes to weigh his options. A man like him knows he is probably a dead man. If he rats and he wrongs the wrong person, well, he is as good as dead in prison as he is on the outside. He will tell us. I know it."

Anne and the officer grabbed some coffee. It was late, or early depending on how you looked at it. It was nearly sunrise. Anne carried an extra cup into the interrogation room.

"Coffee?" She slid the cup across the table to him, uncuffing his left hand.

"Listen, Paul," She leaned in closer to the big man. "I know what you are thinking. You are weighing your options here. If you play your hand right, you could possibly walk free, be able to start over somewhere else. If you play your hand wrong here, chances are you won't be around much longer. I think you know what you need to do here. We can get you a court-appointed lawyer if you want. Or I am prepared to offer you a deal."

"No lawyers. I hate lawyers. They are all liars and scam artists. What kind of deal?"

"You tell me everything you know about the person who hired you. And, in exchange, you go free, with a ride to the airport and plane ticket to anywhere you name."

He sat back in the metal chair. He barely fit on it.

"Anywhere?"

"Anywhere."

"Deal."

HOME, NOT-SO-SWEET HOME

By the time I got home from breakfast, I was aware enough to be looking over my shoulder. I noticed the car parked out front. I nodded to the driver. My dad came in for a few minutes to see Clara. He was trying to keep it light, focusing on me getting through graduation, and the super fun party they would throw this weekend. He tried to act like nothing out of the ordinary was happening when, in reality, there was some sort of undercover cop out front, and everything around me was swirling in disarray. Dad brought in his to-go coffee mug to refill it for the drive home.

Most of my house crew was in there. Sherp and Nik were at the island, eating toast and getting ready to head out for a climb. Nik only had one final left on Thursday and then he was done. Lucky guy. Clara and Marcel were sitting at the small table in the corner with their laptops open, hunkered over them, feverishly typing away. No sign of Carter, which was a good thing for me right now. Kiki must have already headed out to her lab.

"Hey, guys, um, I think most of you have met my dad before. Dad, meet everyone." I did a quick round of introductions.

I stared at Sherp to not even think about laying into my dad regarding the big open question about all of the Brazil connections from last Sunday's family dinner. Sherp had zero filter. He was a tell-it-like-it-is character.

I shook my head at him and hoped he understood that I meant it.

Clara was already on her feet, virtually floating across the floor like she does to give my dad a big hug. I filled my dad's mug with the dark black pot of goo that I guessed Clara had made. It was her style: chewy coffee.

"Yay, Mr. Martin! Wow, what brings you to town so early on a Tuesday?"

"Just a little work, and you know, checking in on my star student during her last week of college."

"Well, Niles, I want you to meet Marcel again. I am sure Meems told you, but Marcel and I are both moving to New York. Hitting the road in just a few weeks, which is crazy to think about."

"Clara, I am so incredibly proud of you. What a great honor to get that fellowship. And you will love New York. I can't wait to hear more about it. For now, I have to run. I just wanted to drop in and say good luck for your final week. And nice to see you again, Marcel."

My dad turned and flashed a smile to the group.

"Nice to see you, kids. Looking forward to seeing all of you again this weekend. Best of luck." We walked to the front door. Dad wrapped me in one of the biggest hugs of my life.

"Just keep your eyes open, kiddo. Trust me, keep your nose in the books and out of this inquisition you were starting down."

I stood back, looked at him, and nodded. "Bye, Dad. Love you."

"Love you, too, sunshine."

My mind was already racing to all of the things I wanted to know but did not, still trying to make sense of this craziness.

HAPPENINGS AT THE FARM

"You need to make this go away. Now! Understand?" Ahmad paced back and forth. Paul had not returned any of the calls and the short man, Stumpy, was nowhere to be found. Ronnie the Muscle and the tall man sat with Ahmad in his spacious house that sat amongst the rolling green hills of the horse farm.

"I think we are good for a little bit, boss. The last shipment already came in and got distributed. Those two idiots probably are running scared for their lives because they couldn't figure out how to take out the bartender," said Ronnie.

"It does not matter. If our bartender wakes up and starts talking, then we have problems. I need to know the job is done, or we need leverage."

"What kind of leverage? That guy has nothing. That is why we knew he would be an easy target. His brother is gone, his parents are gone, he is a loner from Miami in a small town. What could we possibly use as leverage against him?" said the tall man.

"That is not my job to figure out, you idiots. That is *your* job. Surely there is something. And you need to figure out what it is, and deal with it. If we cannot get to him, we need to make sure that when he wakes up, he knows he better keep his mouth shut. Do you understand?"

"Of course, boss," Ronnie said as he stood up.

"Find those two bozos who think they can run out on me and make sure they never run out on anyone else again," he said, resting his pointer finger right in the middle of Ronnie the Muscle's chest.

Ahmad turned, eyes blazing at the tall man who was dressed in a business suit, a Saudi Arabian flag proudly displayed on his lapel. "They were supposed to handle the transit on the next shipment that is set to arrive on Friday. Find someone else to do it."

"Got it."

Ahmad stormed out of the room.

Ronnie and the tall man walked out of the living room and onto the terrace. This was not a good situation. The key business partner in the arrangement was arriving in just a few days. The last thing they needed was any trouble before he got here.

"Do you think I can get in there to just finish the job?" Ronnie said, looking out over the hills.

"Ronnie, you idiot. You would stick out like a sore thumb. I think we need to focus on the leverage for now until the kid is either dead or released, and then we can deal with him."

"You know, Paul did say something about some girl being at the Club and messing everything up. He had that wicked broken nose. I wonder if she did that." He let out a short snort. "He said she was some tiny chick who was there for a little action."

"Well, I think it is worth looking into. There might be something there. Search his apartment for clues and see what you can find," said the tall man. "Find out what you can about her. See if it could work."

Inside the house, Ahmad retreated to his office on the second floor. It was a protected room with a secure line. He made a few calls to his international partners to see how things were going. He had heard rumblings of problems in South America, but nothing concrete. He ran through his list of items to cover with his boss this weekend. Everything had to be perfect. Saudi hospitality was important. The last visit from his boss' team went incredibly well. He showed them around, treated them to a day at the races, hosted a well-attended political dinner at his home. The stakes were high, and the political fallout of anything going wrong would not bode well for him or his family.

FINAL FINALS

I sat in the kitchen nursing a cup of tea and mindlessly staring at my flashcards. It was a routine motion. I was good at routines. Even if I wasn't reading the words on the cards at all, moving them from one stack to the other felt normal. My last two finals were coming up and I honestly couldn't care less. I was still a zombie from last night.

Clara came into the kitchen and grabbed a cup of water.

"Clara?"

"Heya, Meems. Studying hard, I see."

"Hardly studying is more like it. I don't even know anymore. I don't think I am meant to go to law school, but I don't know what I am supposed to do. All of this stuff lately swirling around, it's like a mirror has shattered and all of the fragments are showing me different angles of the same thing but none of it makes sense."

"So it is *that* kind of day. I think you need a beer instead of that tea. We always think better over a beer."

"Yeah, you know what? Let's do this." I went to the fridge and pulled out two bottles of Fat Tire, our new favorite beer that Sherp brought back from his latest climbing trip to Asheville.

"You're worried about Damian, aren't you?"

"Of course, I'm worried about Damian. How could I

not be? I know we have only known each other for like three weeks and that we had called it quits already, but being there when it all happened, it's just too much. I cannot get the scene of the gun being pointed at him out of my head. I keep replaying it over and over, wondering if I should have done something different besides run away. I feel like such a coward. I mean, what if he would have died? I would have spent the rest of my life wondering if I could have saved him. It is so messed up."

"I know the feeling, honestly. I still beat myself up for putting Marcel in a situation where he could have died, where he got shot. It is absolutely crazy that both of our dudes got shot this year, and like only a month apart from each other. What in the hell is the world coming to?" We both paused and took long drinks out of our bottles.

"Clara, I feel like this is all adult stuff and I am not ready to be an adult yet."

"Amen to that. We have about a total of four days left until we are hit in the face with the reality of the world, stripped of the cozy protection and excuse of our college student status. You know what that means, Meems? We need to live it up! For real, this is like our last time to go big as college kids. I think we should go out on Thursday after we both finish our finals. Neither of us have any on Friday, so we have a whole day to recover before the big ceremony."

"Clara, I love you. And no thanks to the night of debauchery. The last thing I need right now is to get wasted and become an absolute emotional trainwreck and lose my shit. But I will go out with you and make sure you get home safe. I haven't seen any good music in forever, and that is my lifeblood. Maybe that is why I feel so out of my body; I haven't been feeding my soul. I cannot go to the Club, though. I'm not sure I can ever go back there. Let's see what's playing at the Elevation."

"Hell yes. To live music feeding our souls!" She tipped her bottle towards mine and we clanked them together. Then Carter entered the room.

"Oh, toasting before dinner. What is the occasion?" he asked.

"Just a little planning to feed the soul cup with some live music on Thursday. You should join us," Clara said, smiling at Carter and then turning to me with a little wink.

"I will think about it." Carter grabbed his water bottle and left the room.

"Clara," I whispered under my breath. "I am not ready for anything right now. So, please, don't push this."

"Okay, I understand. Meems, you are going to get through whatever this is. That dark exterior of yours is filled with rainbows and unicorns, I know it and you know it. Your light needs to shine through. Give it the space to do that. Once you open your heart, the good starts to come to you. Stop boxing yourself into what you think you are supposed to do, or what you are supposed to be. You are the smartest person I know, literally. Your memory and recall are ridiculous, and you just seem to have this instinct about you. Figure out what feels right that makes you feel good. If that is law school, then go for it. If that is Carter, go for it. If it is something else, then open your heart to that."

"Yeah, yeah, yeah." I laughed at Clara, as she has given me some semblance of this pep talk a dozen times over the years. "My unicorns wear black eyeliner."

"And that is exactly why I love your unicorns. Back to studying. I am upstairs if you need anything at all." She slugged down the last sip of beer, took our bottles to the recycling bin, and then headed out of the kitchen, humming *Somewhere Over the Rainbow.*

PUZZLE

David and Anne sat at a computer, virtual conferencing with their team in Washington DC. It seemed the pieces of the puzzle were starting to come together, especially with the information provided by Paul yesterday after his failed attempt at murdering Damian.

They now had confirmation that Ahmad Khan was indeed involved with what was going on locally, and likely what was happening on the international front. To what extent, they still needed more information and proof. They did have a testimony that he was behind the recently seized shipment of illegal timber, where they had also uncovered women and girls in the same shipment.

"This guy is a total piece of shit who deserves to rot in prison. Anyone who deals with trafficking women and children should be shot immediately," said one of the older men on the screen.

"Well, we have a testimony, and I think enough evidence to tie him to it. But we need to see if there is something more. We don't want to eat the bait fish when the bigger fish are still out there swimming," said Anne.

David chimed in, "We have reason to believe he may be hosting a very important guest this weekend. One who may have business ties to the broader network. The coin I got in Brazil. The Chinese agent we captured. The intel from my source in Emirates. I think if we do this right, we may have a chance to bring down the entire organization, or at least a big part of it."

On the screen, a woman stood with her arms folded. She pulled her glasses off and leaned in towards the camera. "Tell us what else you need and it will be there by tomorrow. Assemble the team and make the plan. Anne, you're the lead on this. Don't let me down."

"Copy that. You will have our list within the hour."

PARTY TIME

The next few days flew by. I heard from Anne that Damian was moved to a new room and was stable but still in a coma. I wanted to go see him but couldn't bring myself to do so. Plus, I still had this unsettled feeling in my stomach, like what if it wasn't safe to be around him? Maybe he was right when he said I should stay clear. I promised myself that after I got through graduation, I would go. Maybe I'd take Anne with me for protection. I don't know why the thought of seeing him was so paralyzing. I just needed to get through the next few days and then hopefully everything would start turning around.

Clara blared My Morning Jacket in the bathroom while she got ready. Apparently, she somehow managed to snag us two tickets to see a top-secret show where members of the band were playing under an alias. I don't know how she does it.

I was ready in my usual uniform of black jeans, t-shirt and a black leather jacket, coupled with my ever-steady Doc Martens.

"So we should probably get going if we want to have anywhere to stand."

"Roger that." She clicked off the music and popped out of the bathroom. She looked stunning.

It had been forever since just the two of us went out. It felt like old times. I surprised myself by how much I was looking forward to it.

"Off we go!"

And then my phone buzzed. It was my dad checking in. Crap. I realized I had forgotten to charge my phone this morning. My battery only had a few hours of juice left. I decided to save the battery I had left and call him later.

The line to get in was already wrapped around the corner when we arrived, but it was a beautiful night, so it didn't matter. By the time we got in and got drinks, the stage was set and the lights went dark. We found a place to stand on the top viewing platform, an impossible place to get in and out of, but it provided a great view of the stage. The show was epic. Clara somehow managed to find her way to and from our impossible spot multiple times, this last time bringing back shots for us both.

"Clara, I told you, no one wants to see drunk Mimi tonight."

"So, don't get drunk then. Just do this shot with me to celebrate the fact that we just finished our final finals. We are done. Can you believe it?"

"Well, how could I say no to that. To finishing our final finals!" We cheered and drained the glasses.

The music was incredible. Jim James, the lead singer, had gone to school here for a bit and was a Kentucky native. People here loved him. I had been a fan since high school. The performance was always unpredictable, which is part of the appeal. I hated to step away from it, but I had to.

"I am going to find the bathroom, back soon."

The place was crazy busy. I sandwiched myself between people all the way down the stairs. The bathrooms were down a long dark hallway. Bars like to do this so you can't see how disgusting their floors are. I stepped around two kids who were making out just outside the first bathroom door. I must have startled them because they quickly collected themselves and turned to go back to the concert. I was just about to open the second door when I thought I heard my name.

I turned to look and a guy quickly appeared next to me. As I reached for the bathroom door, a massive hand met my tiny wrist. The random guy flashed me a big smile and stepped in between me and where the lovebirds had been necking. He pulled back his jacket, showing me that he was serious, if the black object protruding from his waistline was what I thought it was.

"Mimi, good to see you. I have been looking for you. Let's take a walk." He led me down the dark hallway to a back door exit.

And just like that, I realized that I was not safe at all, even in this place, with all of these people, far away from the Club. I pulled loose from his grip and turned to run, but he had fast reflexes and grabbed the back of my hair and spun me around, almost lifting me off the ground. I threw an elbow jab behind me to no effect.

"Let's go out back for this conversation," he said as he loosened his grip on my hair.

I quickly spun and round kicked him to the side of his head. I felt my Doc Marten connect. As I landed both feet on the ground and went to run, he said in a deep voice, "I would not run if I were you. I know where to find you. I know where to find your parents. I know where to find your roommates. I know everything about you."

I turned around to see he had the gun out. His eye was bleeding and swollen already.

He motioned with his gun towards the door. "Now, let's go out back."

I froze. Did he really know all about me or was he just saying that? I didn't want to find out. I fidgeted around a bit, trying to decide my next move. If I ran into the bar and he opened fire in there and anyone got hurt, I would have to live with that for the rest of my life.

I walked towards the door. He turned to follow me, the gun pushing into my right kidney, his left-hand bracing on my shoulder.

WE HAVE A SITUATION

"I am telling you; she went to the bathroom and then never came back." Clara was hysterical on the phone with Carter. "I looked everywhere for her. She left right before set break. We had just taken a shot together. She was still sober. She just went to go pee. She would never just leave me somewhere. Something has happened to her, I know it. I just don't know what."

"I am on my way to pick you up. I will drive along the walk route there to see if she just decided to cut out and walk home. I know she has had a lot on her mind. I will be there in 15 minutes, and we can figure it out. Just stay put, okay?" Carter said and hung up the phone.

He drove through the neighborhoods and then down the main drag looking for her. Not many people were out, it was finals week and a Thursday. It would be easy to spot her. But there was no sign of an Asian girl in all black walking the streets alone. He pulled up in front of Elevation. The show was going strong; he could hear it from inside his car. Clara was out front.

"I have walked every inch of this bar and she is not in there."

"Okay, I will talk to the bouncer. Maybe he has seen her."

"Already done that. Not since we came in."

"Where would she go?"

"She wouldn't *go* anywhere! This was our girl's night

out. I think I need to call her dad."

"Don't call him yet. Give me your ticket. I will go in and look for her, too. You stay here, right here, and keep your phone on. I will be right back."

Carter walked slowly through the crowd. He had the benefit of being tall, so he could see over the heads of most people. He combed every inch of the first floor with extreme difficulty, given all of the bodies moving in unison to the song and the drunk idiots falling all over themselves. He took the front stairs to up top. The balcony rimmed the lower floor. She wasn't up here. He exited down the back stairs. That is when he saw the hallway to the bathrooms. He checked each of them, no sign. As he turned around, he saw something out of the corner of his eye. A bracelet on the ground. He would know that bracelet anywhere. It was a thin leather strap with a silver and mother-of-pearl connector that Mimi wore all the time. He picked it up and walked the long hallway. The door at the end was an emergency exit out to the alley. He almost sprinted back to the front door, knocking people out of his way. He threw the door open.

"Call her dad. Now." He held up the bracelet.

As they stood there, a man approached them. He had been sitting in the car out front.

"Clara," he said.

Clara felt like she had seen this guy before, but she was not sure where. She barely looked at him as she dialed Niles.

"Hold on. Put the phone down a second, please. Where is Mimi?" he said as he stepped closer.

Carter advanced. "And you are?"

"It's complicated but, basically, the FBI assigned me to keep watch over Mimi. I watched Clara and Mimi go into the bar, and now I am wondering why you all look terrified and frantic and there is no sign of Mimi."

"She has disappeared. I think something happened." Clara was in tears.

The man had his phone out and dialed a number before she could say anything else.

"We have a situation." He quickly shared the status of what was happening.

"Is this really happening right now?" Carter asked, gravely concerned.

"Carter, this is happening. Mimi had this gut feeling that something wasn't right. She told me. But I just thought she was being Mimi. God, after what happened with Damian, I hope she is alright," Clara said as she walked in circles. She twisted the ring on her index finger as she made the loops, mumbling under her breath.

"Wait, what happened with Damian?" asked Carter. He realized he had been a bit distant this week after she skipped their date. He had been wanting to check in on her but thought she had wanted him to back off. He spent most of the week at the library working on his thesis, trying to stay as far away from the house as possible.

"The team should be here shortly. For now, you two are getting into my car and staying there. Let's go have a chat until the others show up. Tell me everything you remember about the moments before you last saw

Mimi."

About 90 minutes had passed since Clara had last seen Mimi. The show must have just ended because the crowds streamed out of the bar. Carter watched every person exit, hoping he had missed Mimi in the crowd.

Two black SUVs pulled up into the fire lane of the parking lot. Hank climbed out of one, another agent out of the other. The man in the front seat exited to meet them. They motioned to Clara and Carter to get into the front SUV.

Mimi's guard and Hank entered the bar to survey the scene. Carter told them where he found the bracelet and about the back exit.

They were back to the car in about ten minutes.

Clara's phone rang. It was Niles. Apparently, when she had started to call him it went through and then the line dropped. He knew something was wrong. Clara struggled to control her sobbing to tell him what was happening. Her best friend, his daughter, had just disappeared without a trace.

Niles spoke calmly, "Clara, Mimi is a strong and smart girl. Whatever is happening, I believe in her. I have to believe she will be okay. Just do whatever the officers ask you to do. I will be up there soon."

Niles hands shook as he exited the call. How could he tell Hae? He stood up to leave the bedroom just as she entered.

"Niles. Niles, what is going on? What is wrong?"

"I was wrong, Hae. I was wrong thinking that she would be safe. Hae, someone has Mimi. And we need to get to Lexington to figure out who and why."

WHERE AM I?

The bag over my head made it difficult to breathe, especially for the non-panic deep magic breaths that I was trying to engage right now. The gun terrified me enough to know I better do what this guy wanted. I guess I didn't think or realize he would be taking me somewhere other than behind the bar to ask me some questions. That was obviously very stupid of me.

The car ride seemed to have gone on for about twenty minutes so far, maybe forty. It was hard to judge. We were driving fast enough that the car hummed for a while and then turned onto roads curvy enough that my body slid across the seat.

I tried asking questions but was rebuked with a firm warning to stay quiet. I didn't want to see the consequence if I did not. I was laying down in the backseat with my hands tied. At least he hadn't gagged me. I had struggled against him when he started tying my hands. But he was so powerful that it did not matter. But now, he couldn't do much from the driver's seat.

Clara was right. Unicorns were badass. I was going to be a unicorn.

"Um, I have to go pee, really badly. Remember, you intercepted me on my way to the bathroom? I don't know where you are taking me, but if it is much farther, can we just pull over so I can pee? Then you can retie me?"

"Doesn't work like that. Feel free to piss yourself when we get out of the car in about thirty minutes."

"What do you want with me? I have no idea why you would have any interest in me. What did I do?"

"Possibly wrong place wrong time, but more likely you were friends with someone of interest and we need you to get them to do what we want."

"Like what? Do you need me to deliver a message or something?"

"Or something."

The car stopped eventually. The door opened and shut. I heard the sound of walking and then silence. Complete silence. The part about needing to pee had been true. I needed to pee so badly that I didn't think I could hold it any longer.

About ten minutes later, I was dragged out of the car. My feet felt like they were being dragged across dirt. A door creaked open loudly. There was an echo inside the building, it was cold. I felt a wall against my back.

"Have a seat. I'll be back in a few hours. I will bring you some food and water then." He secured my hands to something, I wasn't sure what, then tied off one of my legs to something else. It was incredibly uncomfortable. But I was alive and uninjured.

The door creaked open and then slammed shut. Everything was silent except the sound of my breath and the crickets.

ALL HANDS ON DECK

"Niles, you are too close to this and you know it. The answer is no," Anne said in her calmest, most comforting voice. "We have all hands on deck right now, and some good leads from the bar cameras. So, please, focus on being there for Hae. And keep your phones on at all times in case she calls."

"Easier said than done, Anne, but I understand." Niles looked at Hae as they sat on the same worn couch that Hae and Mimi had sat on only a few weeks ago while they waited for Niles to be rescued.

Hae was wrapped in her long, flowing dress. She shrunk into the furniture and was shaking. The trauma of Niles' kidnapping was still fresh, but now her daughter. Her soul was being ripped out. Mimi and Hae had an unbreakable mother-daughter connection, an intuitive link. Hae could not understand why anyone would want Mimi.

Anne looked at them both. She knew that Hae was not aware of the complexity of this case, or of Niles' real involvement. She shot Niles a look of understanding. He nodded.

Anne left to go meet the team, leaving Niles and Hae at the house with Hank.

"Hank, I think Hae and I should go to Mimi's house. I know Clara is probably going out of her mind. And I worry that if they took Mimi, are they going to come back for Kiki or anyone else?"

"We have two guys stationed over there, Niles. But if you want to go there, we can."

Niles looked at Hae. She was not going to be making any decisions right now. She just wanted her daughter back. Eventually, she shook her head in agreement then collected her purse and phone from the table.

Meanwhile, Anne met up with the rest of the team at their new official office downtown. Too many people to have coming and going from the house, which had transitioned out of being a safe house to being an office house. It was no longer a good base for operations. Several agents had already arrived from DC and other local field offices. This office provided more space and better equipment. They were planning their raid of the Khan property for Saturday evening. This incident complicated the original mission.

"Let's talk about why we think they took Mimi and then where," Anne spoke with authority to the group of ten people sitting in front of her.

David spoke up first, "On the why, Mimi was there when Khan's men came into the bar and shot Damian. She's either a liability because she saw them, or they are using her as an asset for leverage. Either way, she is in danger. But if she is seen as a liability, then our clock is being measured by the minute, not the hour."

"Good. We have everyone looking for the plate of the car that we think she left in."

"Right, but this could be a totally random unconnected event, just circumstances make it seem like they should be connected," said another agent.

"Agreed. We have to keep all options open. Right now, we don't have an ID on the guy she left with, but we do know there was a car in the alley about the same time, so that is what we have to go on at the moment." Anne looked at the whiteboards of information in front of her.

"I think Mimi is in extreme danger and we should treat this situation as such. I think we need to move up the timeline for the raid," David said.

"I concur. I have looked at all of the evidence in this case. This guy, Khan, is a real special guy. Pretends to be perfect with his horse farm, businesses, hosting top politicians and businessmen. If this evidence is right, and he is the person behind this ring of global illegal timber and human trafficking, then he is way smarter than we think. He will have people everywhere. His deals being worth millions of dollars, he will be ruthless. If in fact Mimi's disappearance is linked to this case, which statistically speaking, it likely is given all of the connections, then I agree with David, she is in extreme danger," said the female agent sitting next to David.

"Let's scenario this out. If we go in tonight to get Mimi, assuming she is there at the farm, then we lose the chance to see who he is hosting at his dinner tomorrow. Let's keep working this for now. See if we can check traffic lights and any other cameras in the city to get traces on the car's plates. See if there is any indication of them being somewhere besides the farm. Let's get someone over to interview Mimi's roommates to see if they know anything else. And for the love of

God, let's hope Damian wakes up and can give us more information. I want a report back every 15 minutes on findings. If we have a solid lead, we will go."

Anne's phone pinged.

"They have video footage from within the bar. It is dark and hard to decipher, but it looks like our Mimi is quite the skilled fighter, and that there was a struggle before she was taken. Let's be prepared and hope she is not injured."

Anne pulled David aside. "You are too close to this, too. You are not going to the house. I want you here. I want you to make some calls. See if we can get a strike team ready to go onto the property and scan for indicators, unseen. Let's see if we can find her without this whole thing blowing up on us."

"Copy."

Clara rose to get the door. It was Niles and Hae. She all but fell into Hae's arms, trying to console her best friend's mother while seeking consolation for herself.

"Mrs. Martin, I am so sorry. This is all my fault. She wanted to stay home but I convinced her to go out. I didn't know she was in any danger. I had no idea someone was even watching our house. This is all so messed up."

Kiki stood from the dining room table. She didn't advance, just looked at Niles and Hae, her eyes expressing her sorrow.

Marcel stood next to Clara. He stammered out in a quiet voice, looking Niles in the eyes, "I am so sorry, Mr. and Mrs. Martin."

"Marcel, any chance you all have some tea?"

"Of course, Mr. Martin."

"Two cups please, whatever you have. No milk. We aren't British or anything like that," Niles said, trying to lighten the mood a little. These kids were clearly shaken to the core, as was he.

Niles and Hae walked with Clara to the dining room and sat at the long table. Carter was upstairs, Sherp was at work, and Nik was hunkered down on campus somewhere studying. Clara hadn't thought to reach out to everyone. She was only focused on herself.

They sat there silently while Marcel was in the kitchen getting the tea.

"I take it you haven't heard anything yet?" Kiki finally spoke.

"Not yet, no," Hae responded in a near whisper.

Marcel entered with the tea right as there was a knock at the door. Clara jumped up to run to the door, but Niles grabbed her arm. He knew Hank was outside, standing guard, but he did not want to take any chances. Niles walked slowly to the door, looking out the front window first. Two agents stood at the door. Hank talked with them from his post on the porch.

"May we come in?" they asked Niles. He did not recognize these two. They must have come in long after he left Washington.

"Of course. If you are standing here, that means there is either good news or bad news, and if it were good news, I think Anne would have come to deliver it herself, so just tell us." Niles' voice was steady and calm.

"Actually, no news yet. We came for information. We were hoping to talk with the roommates about anything else they remember, any other possible lead."

Clara repeated what she had told the agents at the site. Carter came down and did the same. Kiki shared that she been home studying.

"Kiki, I am not ruling anything out here. I think you should tell them about Brazil," Clara said, grabbing Kiki's hand across the table.

She acknowledged what Clara was saying and then looked straight at Niles.

"While I was in Curitiba at the conference, I went on a walk at the botanic gardens. While I was on my walk, a man attacked me from behind, but I was able to defend myself and get away. He didn't take my backpack. I don't know if it was a mugging or an attempted rape or what it was, but I got out of Brazil as fast as I could."

"Thank you for telling us. We will note that in the file. Anything else?"

Kiki continued, "Is no one thinking what I am thinking? Maybe this is obvious, but Meems had recently broke it off with a guy she was dating. He owned the Club, right on the main strip. I think they had only gone on a few dates in the last few weeks, but still, I wonder if he could be involved somehow."

"To be clear, Damian and Mimi are not a thing. At all. It was just a few dates, nothing more, based on what she told me. She did not make it seem like there was any bad blood there," Carter chimed in.

"Plus, Damian would never do this. At least, I don't think he would. And how could he? He is in the hospital," said Clara, defensively protecting him for some reason.

They all looked at her. Apparently, Mimi did not share much with her roommates about her personal life.

The one agent asked a few more questions while the other agent went with Niles to go through Mimi's room for any other possible clues as to where she might be. They did not turn up anything exciting, but then Niles spotted Mimi's MacBook on her desk.

"Well, I'll be damned," he said.

The agent turned to look.

"Find my iPhone. Come on, sweetheart, have it turned on, please." Niles flipped open the lid to her computer only to be blocked by the password box. He flipped the laptop over, searched around her desk, nothing.

He yelled down the stairs, "Clara, can you come here, please?"

She sprinted up the stairs, followed by everyone else.

"Clara, do you have any idea what Mimi's computer password is?"

"Um, no, I know her phone password though. I guess it is worth a shot. 555528," she typed it in as she spoke.

Incorrect.

"Think, think, think," Niles talked out loud to himself. "Hae, love, we have to think like Mimi. What would she make her password?"

Carter walked around the room and picked up a notebook next to Mimi's bed. The agent had already looked through it on the first search of the room. Mostly phone numbers, to-do lists, reminders. Carter flipped to the back page. Random words all the way down.

"Try ClearanceClarence1980."

Niles smiled broadly at the suggestion. *Of course,* he thought to himself.

"Bingo, we are in. Where is 'Find My' on this thing?"

The agent already had Anne on the phone. She put them on speaker in the new office so the other agents could hear the update.

"The location looks like it is in the opposite direction of the horse farm. Jackson County, near McKee. Sending pin drop for the location now. We will meet you on the way out of town." All of this was being said while they headed down the stairs and out the door, laptop under Niles' arm.

"Put Niles on the phone," said Anne to the agent.

"Here, Anne."

"Stay put with Hae and the kids. We have this. I need you to trust us. I cannot have a civilian with us on this."

"No way, Anne. This is my daughter. I am coming."

"Exactly, Niles, you are way too close to this to make good decisions. We will be in touch soon. We will take care of your girl, I promise."

She ended the call. Niles handed it back to the agent whose steps quickened down the path to the car.

Niles, Hae, and Clara were left standing on the porch with Hank, who was assigned to protect the house, and another agent who stood by in a car down the street.

McKee was about an hour from where they were if you drove normally. In a police car, maybe 45 minutes, given the road conditions.

They headed out to Richmond Road, blowing through traffic lights. They got to the rendezvous point in about 15 minutes and waited about five more for their leader and the others to show up. When they did, the four SUVs flew down the four-lane highway to the interstate.

"I have people running satellite photos of the pin drop for us now. It doesn't look like her location has moved at all, so hopefully that means she is being held somewhere and they are done moving."

The phone rang through the car's Bluetooth.

"Rodriguez," she answered.

"This site looks really interesting. One main building and three auxiliaries. A pretty big area surrounded by woods. Looks semi-secure with fencing and a gate in the front. It is hard to know where she is within that area. Those GPS locators are only good for up to about 20 meters, and that is if there isn't any other disruptor in place."

"Got it. Call me back once you get intel on those aux sites and if we are looking at any issues with security." Anne ended the call.

"Three teams. One for the perimeter, one for auxiliary buildings, and one for the main building. David, you and I are going to run the show from out on the road."

Anne grabbed the radio. "Listen up. We are 20 minutes out. Get prepared. Maps being shared now."

Anne looked at the maps and knew exactly the location Mimi was being held. They had her at the mill site where Marcel was shot. It was clear now that the mill was some sort of primary hub for activity. She made a mental note of that and then ran through strategy with the teams. To not trigger possible security at the site, two SUVs would park at the side road and walk down, one would park at what looked like a pullout on the map, and one would go past the site and park on the road about 100 meters up. Everyone looked at the satellite photos that had been sent and identified access points to the site and the buildings.

"Let's bring Mimi home."

Over the secure channel, the teams called when they pulled into position. They moved silently out of the cars, ensuring they were fully equipped. Anne spoke through the earpieces.

"Team One."

"Check, ready."

"Team Two."

"Check, ready."

"Team Three."

"Check, ready."

"It's go time. Team One advance and report on the perimeter security. Identify best access."

"Back gate access. The two cameras are down. Perimeter secure."

"Team Two and Three, go."

"Copy, moving now."

About two minutes passed.

"From out, it doesn't look like anyone is home. Quiet and dark on the perimeter."

"Team Two entering main building now."

"Team Three entering auxiliary building one now."

The agents took out cameras as they went. They had disabled the power before entering the buildings, so the cameras should have already been down unless they were on a backup generator. It seemed any security lights in the area were already out. It was pitch black. The air was completely still. It was humid. You could smell the resin from the fresh cut wood. Logs were stacked ready to be fed through the equipment. Large piles of chips, about 25- or 30-feet tall rose from the ground. The agents navigated around the debris and equipment in the large warehouse in silence. So far, nothing.

"Aux one clear, moving to aux two."

Anne looked at David. He watched the teams through their goggle cameras on his laptop.

"Aux two clear, just a tiny electrical building. Moving to aux three."

Team Two continued their methodological walk through the warehouse. There was no sign that anything was out of place.

David jumped out of his seat. "There she is. There she is." David looked at Mimi sitting in an awful position with the bag over her head, but she was moving. She clearly heard the footsteps.

"Confirm you have a visual on the target."

"Confirmed, target alone in aux three."

"Team Two, retreat to position."

"Get away from me. You are going to be sorry. I promise you. You fucker. Not even letting me go to the bathroom. What do you want with me anyway?!" Mimi screamed at the team approached her.

"Mimi Martin, you are safe. We are here to rescue you. You are no longer in danger." The agent pulled the bag off of Mimi's head and cut the ties from her hand and foot.

"Are you injured?"

"No, I don't think so. Just really thirsty." Mimi rubbed her eyes and her hands. She tried to stand but her leg was asleep and her back stiff.

"Let's get you out of here before someone comes back looking for you."

"Team One has secured uninjured target. Moving to rendezvous."

The other agent picked her up and carried her as she clung to his neck. He moved silently and gracefully through the darkness towards the forest. Within minutes, she was being placed in the back of a car and covered in a blanket. The car drove down the dark road without headlights on for several miles before turning onto what looked like a rural two-lane road. The female agent handed a bottle of water and a granola bar to the back seat.

"We are going to get you checked out once we get to where we are going. Are you all right for now?"

And the head nod confirmed yes.

Anne and David were at the rendezvous. Team Two

stayed behind at the facility to see who it was that would be coming back for Mimi. Anne had already made the call earlier today to have Damian moved out of the hospital into private care. Two more agents were assigned to Mimi's house when they drove off for the mission. She gave David permission to text Niles. *We have her, safe and sound, and uninjured. She is okay.*

Anne called Hank with instructions to take the kids' cellphones and computers earlier that night. The last thing they needed was any social media mishap or leak about the mission. She wanted them all to stay put until this thing was over.

David looked at Anne. "We have a few working theories on motive, but nothing solid. Most obvious is that they realized she was the girl with Damian when he was shot, and they wanted to silence the evidence."

"That makes sense. Let's see if she has any ideas. Find out who she talked to and what she heard."

David looked at Anne deeply. "Anne, I know we have to question her, but she is fragile, and Niles' daughter. It goes without saying that we need to take extra caution with her."

Anne nodded in acknowledgment as her phone pinged.

"Even better, Damian is waking up out of his medical coma, so we should have more answers soon."

UH-OH SPAGHETTIO

Ronnie had put the girl in a safe place and then went back to town to take care of some unfinished business and to run a few errands. He let the boss know that he had gotten the leverage they needed, and he had it secured. Remote, middle of nowhere, closed for the weekend; they had at least 48 hours to use the leverage in its current location. He was keeping her out of the way until after the weekend guests left.

He headed back out of town and on his way stopped at Walmart for a blanket and some bottles of water. There was a McDonald's in the parking lot. His stomach growled. He hadn't eaten all day except for that old protein bar he found under the seat. He whipped the car through the drive-through. Two quarter pounders with cheese and a chocolate shake. He paused for a second and then added on a cheeseburger, fries and a Coke for the girl. He felt kind of bad leaving her there like that for the last few hours, but she needed to know he wasn't messing around. Maybe she would shut up and stop asking questions now that he had taught her a lesson.

As he exited the drive-through, he got a ping that the guest's flight had landed at the Lexington airport. They were spending tonight with family then tomorrow was the big dinner party. He still had plenty of time to finish all of the jobs the boss had given him. Most importantly, silencing any dissidence that might be brewing.

He liked driving the dark winding roads and taking the turns fast. It reminded him of another time, another life he had led before landing in this backwards little city. He slowed as he approached the mill, scanning his key card for the fence gates to open. He drove to the back of the lot, to the small building near the rear of the property. He slammed the car door shut behind him and clomped towards the building. He regretted dismantling the security lights earlier as he could not see a thing, but he knew this place well.

He fumbled with the door, finally getting it opening and announced, "I hope you are hungry. I brought you McDonald's finest," as he clicked on the light.

It took his eyes a moment to react to what was in front of him, which was absolutely nothing except for the rope he had used to tie up her leg. He stumbled around the small building looking to see if she was hiding somewhere.

He mumbled under his breath and kicked the dirt. "Fuck, fuck, fuck. I am a dead man."

Ronnie ran out the door towards the main building where the power panel was and flipped the breaker to turn on the security lights while yelling loudly, "I know you are out there, sweetheart. Come on back. I have a nice dinner for you."

He slowly walked, carefully looking around every bend, in all of the outbuildings, and then he noticed the back gate was open. Normally, that was deadbolted shut. She could have easily climbed any of these fences, too. Still, he had to ensure she was not hiding somewhere on the premises first. Who knows how long ago she had broken free? Regardless, out here in these woods, she would not get far. No one drives out here in the middle of the night, so there wouldn't be anyone to pick her up. His mind continued to race as he searched high and low for any sign of her. He finally got back in his car and decided to drive up and down the roads.

When he drove up to the gate, it would not open. He clicked the button three times and still nothing.

"Great, now the damned gate is broken, too." He got out of his car to manually move it.

He suddenly felt his body shake uncontrollably as a tremendous electric pain pulsed through his body and took him out at the knees after several seconds. When he finally realized what was happening, he saw the two men standing over him with a taser and handcuffs. He spewed off a string of profanities and writhed on the ground, trying to find his feet to stand up. This resulted in the one officer kicking him squarely in the kidneys. He may have lost consciousness while they were securing his feet. Before he knew it, he was in the back of an SUV, strapped in tightly and unable to move his arms or legs. A man was sitting next to him.

The other agent had his phone and was connecting it to a computer in the front seat of the car. He made a call, "Coming through now."

"Don't worry. We turned off your tracker. Last known location was right here at this luxurious facility."

Ronnie hung his head. He wished they had just killed him. He was as good as dead now, but his death would be way more painful and drawn out than if the officers had just shot him.

ONE MORE SATURDAY NIGHT

I was so exhausted but was finally in my own bed. The first rays of sun were beginning to fall on the big tree outside my window.

Yesterday had been the longest day of my life. The interrogation after the rescue seemed as painful as the abduction. I know they meant well, but my head was swimming, and I didn't know the answers to all of their questions. I sat there, in the purple chair, doing my best, but I just wanted to sleep.

At least it wasn't like in the movies where they take people to some cold room with a metal table and chair and a single incandescent light. My mom and dad were there by my side. They let me shower and change my clothes before they asked any questions. And they were super patient with me. I was just so confused about what had happened and why. I needed to be done, and my dad agreed, letting them know they could finish any questions they had tomorrow, and that he was taking me home. I fell asleep thinking how grateful I was to get a day of rest before I would walk down the aisle in my commencement ceremony.

Little did I know that while I was sleeping there was a massive debate happening about whether or not I *could* go to graduation.

I couldn't even fathom the fact that I was supposed to walk in a graduation ceremony tomorrow afternoon.

I had the whole day today to rest and recover, but I just didn't think I could bring myself to do it after everything that has happened.

When I finally came downstairs it was the middle of the afternoon. The old wooden stairs creaked under my foot, causing Marcel, Clara and Sherp, who were sitting on the couch, to turn and look at me. I wish I could avoid everyone right now. The last thing I wanted to do was play another round of twenty questions.

Clara leapt to her feet and wrapped me in a warm hug the minute my feet touched the landing. I patted her nicely on the back and said thanks.

I turned to walk towards the kitchen. I felt like quite the zombie. I was really out of it. That did not stop me from noticing the beautiful bouquet of flowers on the middle of the dining room table. I glanced at the card. Clara's dad had sent them to her.

I fumbled my way to the kitchen, half asleep, to find something to eat. Out the back window, I saw a strange man standing there and immediately went into panic mode. I turned to run to the other room and bumped right into Carter.

"Run, they are back. Run!" I tried to push him out of the way but he wouldn't move. He just wrapped me up in his arms.

"You are safe. Those are the police. They are watching the house to make sure everything is okay. You are safe. Breathe," he said, softly stroking my hair as I melted against his chest.

The panic had taken every ounce of energy I had. I didn't realize I would react the way I did. I clearly am going to have some stuff to work on if I ever get back to the therapist.

I didn't know where the sound was coming from, the raspy breathing to a near choking. Then I realized that I was uncontrollably sobbing against his chest. He just held me and stroked my hair. When I finally settled down and the shuddering in my chest subsided, he relinquished his grip on me slowly. I think he was worried I would fall right to the floor if he let go.

He grabbed my hand and led me over to the table. I sat down. He went to the fridge and moved about the kitchen without words. Within five minutes, I had a plate of food in front of me and a cup of hot tea.

"Do you want me to stay?"

I nodded and he sat across, silent, just looking out the window.

I tried to eat a few bites but lost my appetite. I sunk into the bench seat with the tea and followed his gaze. A monarch butterfly was perched on the edge of the hummingbird feeder outside the window. Carter had an easy smile on his face while he looked in wonder at the sight.

"Did you know that there is a forest in Mexico where most of the monarchs go over winter? They fly all the way from here to Mexico. Amazing that this tiny creature can travel that distance," he said.

Carter continued, "They are magical creatures. I mean, metamorphosis, really? One day it's a caterpillar, and then another day, it emerges this beautiful, winged being."

I pushed myself out from the bench seat and walked

closer to the window. The butterfly's proboscis unraveled into the tray below the feeder as it lightly opened and closed its wings. I felt it looking at me then it gracefully lifted off the tray and moved to one of the lilac bushes in the yard and then onto the rhododendron. It seemed so carefree, just observing the beauty of the yard, trying to decide where it would go next on its journey.

I turned around and Carter was still at the table. I dropped my chin and let my hair fall over my face. I didn't want him to see my puffy eyes. I slipped back into my seat and took a few bites of the sandwich.

I am not even sure the words came out of my mouth, but I felt like I said, "Where in Mexico?"

"I'm not sure, but I did hear that there is a legend that the butterflies are the souls of those who have gone before us. In Mexico, many consider them sacred. That seems about right. Every time I see one, it takes my breath away."

We sat there together, silently, for what could have been an hour.

"Do you want to go back to bed?"

"I think I probably should. But I am worried that when I close my eyes, the darkness will come and I will be back under the bag, tied up in the car and left for dead in a shack in the middle of nowhere. So, I don't think I could go back to bed if I wanted to."

"How about I grab a book and come sit in your room? If I am there, nothing can happen. You can rest and I will be there if you need anything."

"Thank you, Carter."

When I woke again, he was gone, and my mom and dad were sitting there. As soon as my mom saw me stir, she was at the edge of the bed stroking my hair. I looked at the clock and it was nearly six in the evening. I had slept way longer than I realized.

"How are you feeling, kiddo?"

"I have had better days, Dad. But at least I am okay."

"The only thing that matters is that you are alive and we are all here together. Hank just went to pick up Thai food for everyone, if you feel up for coming down to eat. No pressure. You can eat up here, too. Whatever you want."

"Wait, are all of my roommates still trapped in the house? Are we still in danger?" I felt the anxiety creeping back.

"No, honey, they are just being extra cautious. They caught the guy who abducted you and are questioning him. They are just asking everyone to hang tight until that is done," my dad said.

"Dad, tomorrow is graduation. We all have stuff to do to get ready. We can't hold them here like hostages."

"If they need to leave, they can. No one is being held here against their will."

I took a deep breath and sat up nodding that I understood. "And Thai food sounds great. I'll come eat with everyone."

I looked at my mom and dad and realized this had aged them. My mother, normally so composed and perfectly coifed, looked exhausted and ragged. My father always looked this way, but his eyes showed me how tired he was.

"Dad, is it really so bad that you don't have any terrible jokes to cheer me up?"

"Well, you know, this morning Siri said to me, *Don't call me Shirley*. I must have left my phone on *Airplane* mode."

"Oh, sweet Jesus, Dad. That one . . . wow."

"Well, the only reason we found you was because of Carter finding your password book and us realizing that of course, you would use a reference to that movie as your password, so I thought it was very apropos."

There was a soft knock at the door. It slowly opened and revealed Clara.

My mom and dad stood to leave the room, Clara came in. She paused for a minute until my parents were out of the room and then came over to the side of the bed.

"I am so sorry I made you go out and I took you to that place." Tears fell down her face as she said it.

"Clara, I chose to go with you. You did not make me go anywhere. If those people wanted to get me for leverage, I am glad they did it there versus breaking into our house and possibly hurting some of you all. I am a little traumatized but will be fine."

"I just couldn't believe you were gone and then Carter came to help me look for you and—" I cut her off before she could finish.

"Carter came to look for me?"

"Yeah, of course. I called him and he was incredible. So calm, so organized. He was amazing. I think he must really like you, or he is just a good guy, or both."

"Clara, I just want you to know, because I know you

want to ask me but aren't . . . That guy did nothing to me. I was just a hostage or something. I just want you to know I am okay."

She leaned in and hugged me again. "Oh, thank God."

"Let's go eat."

I came downstairs and it was amazing to see all of my roommates and my parents, and even better, Clara's dad, sitting at the same table. It was like being at a wedding or a funeral or something, where everyone you care about gets together. I guess it *was* graduation weekend.

I tried to be normal, but I knew my gaze was a million miles away. I can now relate this face to where Kiki was a few weeks ago, just dazed. I was sitting next to her. She reached under the table and gave my hand a squeeze. My mom was on the other side of me. Dad kept having to step away to take calls.

After dinner, he and one of the agents sat on the porch in a very intense conversation which made me incredibly nervous. But my mom, Clara's dad, Clara, and I sat in the living room watching *Lucifer*. Clara told them about her final project and her big internship as well as her and Marcel's upcoming move. Clara's dad caught Mom up on the world of beer brewing and his latest homebrew invention, offering her a sneak sample, which she gracefully declined. I kept telling myself this nightmare was over and everything was normal and safe.

WHERE THE HELL IS RONNIE?

The boss had texted or called six times before 9 a.m. Ronnie hadn't answered. Ahmad cursed under his breath. Ronnie was supposed to be at the house. He had other guys working all of the angles to ensure tonight went seamlessly. Ronnie was his muscle, which he flexed often. He needed to know the status of his problem, and his leverage, and was not getting reports on either, which caused him great concern.

He sent one of his other guys down to the hospital to do a walkthrough and see if he could get a status on Damian, hoping he was still in his coma, at least until he could get this deal closed.

Damian, and the ten others like him that Ahmad had around the world, were going to need to be on the ready to move this money quickly. He could not take the risk that something would go wrong. He knew Ronnie had figured out who the girl was and had her in a safe place. He also had his guys running background on the girl to see if there was any cause for concern with her being involved.

A tall, thin, balding man with short, cropped hair and a pocketed face approached Ahmad, who was sitting outside on the terrace observing the tent being installed in the lawn.

"Sir, we may have an issue. Apparently, the girl's father, well, he is a former federal investigator. We may have picked the wrong asset to leverage."

Ahmad sat with this information for quite some time before speaking. His eyes spoke first as he turned and stood, cup of tea in hand. He was a tall man and a very imposing figure in his traditional clothing. The two others sitting next to him also rose.

"Find Ronnie."

Ahmad's guests were set to begin arriving at 4 p.m. He had arranged for a tour of the farm first to show off his finest thoroughbreds. There was hope to sell some semen, and the stud fees for his best horse were quite lucrative. He had a mare in foal now, about to give birth, as well. He was very proud of this horse and planned to show her off.

After the tour, they would retire to the patio and tents for cocktails and live jazz and the rest of the guest's arrival, followed by dinner and a traditional Saudi performance. He was expecting about 75 distinguished guests for the event, including the president of the university, several well-known up-and-coming politicians, business associates from around the world, and, of course, his guest of honor and his entourage and family.

Around noon, the tall thin man returned. "Sir, no one can reach Ronnie."

"Send someone down to the Beyer's Mill off of Big Ridge. I think he had some business down there this morning and maybe got tied up."

"Yes, sir. And one more thing, sir. The hospital said that Mr. Lopez was moved yesterday to another care facility, but they would not release which one. We have our people on this. We should know soon where he was moved and what his condition is."

Ahmad sat back in his chair, playing out the dozens of scenarios for what could go wrong or was going wrong. He needed to plan for everything. He leaned forward on his elbows and summoned the two men, one on either side of him to lean in, as well. He reviewed his contingency plans. It involved a lot of manpower to protect his event from anything going wrong. They were both on their phones within seconds, working the assignments through their networks.

INFILTRATION

All of the plans were in place. Ronnie had been less than forthcoming with information. He was not as easy to turn as the other hitmen. From his phone, they did know some sort of deal was pending and the names of companies that were providing services for the day. That allowed them to put a woman on the set-up crew for the tents. She infiltrated the property and installed a few bugs. Another agent was placed on the catering crew. They had people on the inside and installed a camera across the street to monitor incoming traffic from the road into the driveway.

The team had put together enough information from Ronnie's phone and Paul's intel to know that at least two ringleaders for the illicit trafficking ring would be in attendance. What they did not know was who the kingpin was and if he was the guest of honor.

Other than some minor activity early in the day, it had been very slow at the house. Ahmad had gone inside after lunch and was not seen again. The tents were up and bugs in place on the exterior. The catering crew arrived at two o'clock to set up. They did not want to risk bugging the house and being found, so the agent wore a wire. It seemed oddly quiet given one of Ahmad's top men had disappeared. That was a little disconcerting.

Two teams would be on standby on adjoining

properties. Anne and David would enter as a couple, claiming to have an invite. This disruption would give the necessary clue for the teams to advance on the targets. Everyone was in their positions. It was four o'clock. Go time.

The first guests arrived at a few minutes after four, followed by about a dozen others. These were the guests invited to the private horse tour. It was difficult to see individuals as they arrived, but as the half dozen golf carts were loaded with guests to go to the horse barn, the agents on the west side of the property had a clear view.

Everything seemed to be going according to plan. There was nothing out of the ordinary.

The third cart was the one of most interest to the agents. The other carts contained well-known horse breeders and owners, as well as a few prominent businesspeople. The third cart contained Ahmad and three Asian men. The agents took photos, quickly sending them off for cross-check in their system. The one in the front must be the guest of honor.

The barn had also been bugged. Nothing out of the ordinary was said so far. The tour was concluding, and the carts were heading back to the house, where other guests had begun to arrive. Jazz music filled the air. Anne was starting to question this raid.

"Stay the course, agents. We don't know what else to expect tonight," Anne spoke into the device built into her watch that transmitted to all of the agents.

"Zeta will enter at eighteen hundred hours as planned. Be diligent."

"Copy," said the agent from Team One out in the western field.

"Team Two, do you copy?"

Silence.

"Team Two, do you copy?"

Anne picked up her cell phone and made a call. "Ma'am, we have a situation. Okay. Yes. Team Three in position. Copy."

She looked at David. "Orders are to proceed, now."

"Zeta mobilizing ahead of schedule. Entering at seventeen thirty hours, in three minutes."

"Copy. Standing by," responded Team One.

Still nothing from Team Two.

Anne and David exchanged a glance and pulled the car into the long driveway up to the house.

Anne stepped out in a stunning red cocktail dress, fitted on the top, flared at the waist with a lovely flow to her knees, with capped shoulders that showed off her athletic figure and flawless dark skin. David was incredibly concerned about her dress choice because it meant she would not wear her full vest, but only the more pedestrian ballistic protection tank top. David wore a traditional black suit, complemented with his salt and pepper beard, looking incredibly distinguished. They held hands as they approached the door, which the butler opened for them.

Inside the door stood a man with a tablet who greeted them.

"Welcome. Names please?"

"Mr. and Mrs. Sheraton Downing of Downing and Associates," David said with his rich deep voice.

"Just a moment." The man turned away from them and typed something into his cellphone.

They had anticipated this and created a fake web profile for the business and the man.

"Just another moment, please. I am sorry, my tablet is acting up and I am not showing you on my list."

"Not a problem. Oh, look, there is Senator Hypkins, love. We do need to go say hello on our way in," David said, doing his best to beam at his fake wife and play this part.

A large core of agents and investigators were stationed just up the road at a church in a van and an SUV listening to everything going on. One group had already been deployed to identify what was happening with Team Two. The agent on the inside had confirmed that whoever the Chinese guests were, they were the center of Khan's attention, clearly the guests of honor.

"I am so sorry, Mr. Downing. Just wait right here. Mr. Khan will be right here to greet you." He stepped back to greet another set of guests.

Anne reached up to scratch her nose and whispered into her watch, "Now."

At that same moment, six large men walked up behind Ahmad Khan as he approached Anne and David. He bore a large grin on his face, hands clasped behind his back, his head slightly cocked to the left as he investigated them.

"Mr. and Mrs. Downing, so nice to make your acquaintance. I don't believe we have had the pleasure of meeting before," Khan said through his smile and took David's hand in his, shaking it firmly.

The six men flanked the agents.

"We could not miss such an important event with such distinguished guests. I have driven by this property many times and admired its beauty. I always wondered what lie behind those stone walls and wrought iron gates. Thank you for having us. It is as interesting inside as I had imagined." David smiled back, staring Ahmad right in the eye, their hands still connected.

Ahmad pulled David in closer while looking over his shoulder directly at Anne, scanning her up and down. "Tell me, Mr. Downing, what is it you hoped to see inside these gates tonight?"

"Well, you, of course."

At that moment there was a commotion out on the lawn. The band stopped playing and people parted like the Red Sea. The large glass doors that opened from the back of the house out to the terrace allowed Anne and David to see what was happening. Ahmad still had David's hand.

"It appears I am needed on my terrace. I apologize for the inconvenience, Mr. and Mrs. Downing, but I am going to have to ask you to leave now. My guest list is full for this event. My men will escort you out."

Ahmad turned, not to walk out to the terrace, but instead he went down the hallway to the left, followed by two of his men. This left four men surrounding Anne and David.

In the thirty seconds it took for this exchange, screaming erupted. Gunshots were fired on the terrace. They were semi-automatic shots, unlikely fired by

Anne's agents. The sound was wrong.

Anne reached for her thigh where her Glock was holstered. As David went for his, the man to the right of him grabbed his arm. The guards had pulled out their guns, pointing them at Anne and David.

Within seconds, the SWAT team came through the front door and took out Ahmad's men before they could get a shot off.

Anne took off running down the hallway with David and another two agents. Each carefully checked rooms as they went. Precious minutes passed. They heard the commotion on the lawn continue.

At the end of the hall was the sprawling main bedroom with three massive closets. They searched but found no sign of Ahmad. Doors from the bedroom exited out to the far end of the terrace, away from the party. Two men opened the French doors and scanned the area for any clues.

"I found something."

Anne entered the largest of the closets to find David standing in front of a small door that had been covered with a large mirror. He'd seen the hidden hinges and knew this type of wall. It led to a dark and narrow passageway.

Anne and David entered. David pulled the tactical light from his vest and held it above his gun.

"We have a possible escape route headed to the east. A tunnel. Zeta entering."

They crept down the dark hallway for several minutes. It appeared to be shipping containers underground framing the tunnel. This must have been a massive excavation project. They came upon an area where several dirty, used mattresses were lying folded up against the wall. Anne and David exchanged a look.

"We have gone about 300 meters," Anne said through the mouthpiece on the watch.

Then they stopped. The large tunnel narrowed and they felt air moving through the area. A ladder went straight up above them, but the tunnel kept going.

David motioned that he would keep going in the tunnel and for Anne to go up. Her dress was no hindrance as she made the first step up the ladder, quickly climbing the rungs.

Anne was forcing open the large metal hatch above her when she heard shots ring out. She heard at least ten rapid bursts from a semi-automatic weapon, followed by the sound of the Glock.

She jumped down off the ladder as quietly as she could and advanced at a near sprint. She did not turn on her flashlight to retain the element of surprise. After about five seconds, she tripped over something on the floor. It was soft and warm, a body. She listened carefully, no breathing, nothing. She held her gun steady while she flipped on her flashlight to see what lay beneath her.

"No, no, no. No!" Sadness and rage filled her. She leaned down over David to find a pulse but knew there would be none. They got him above his vest.

"Agent down. Agent down in the tunnel."

The voice in her ear responded, "Team on the way. Should be arriving in a few seconds." Anne heard the footsteps down the tunnel.

"Proceeding to target."

Anne continued down the narrow hallway. The breeze grew stronger. She knew she was reaching the end. She looked up, another ladder, but this time the hatch was open. She heard the sound of motorbikes in the distance.

"I hear motorcycles. It is possible that the target is escaping by bike." She scanned the area and saw a small shed with the door open and tire tracks on the ground.

"Confirmed. They have left by bike."

She went back to the tunnel to be with her partner. She arrived to have her worst fear confirmed. She spoke a few words to the men handling the body and promptly left back through the tunnel into the house.

By the time she got back out to the terrace, she found terrified people hunkering in corners. The whole operation had lasted less than fifteen minutes. She surveyed the situation. One of the agents approached her, the Special Agent in Charge.

"Four of the targets were secured and are alive. Two are confirmed dead. One agent on Team Two is confirmed dead and one is injured. Six of Khan's men are confirmed dead. Three are missing. No civilian casualties, one minor injury."

At this point the local police were arriving, along with medical teams. A helicopter had been called for the down agent.

"Thank you, Agent Sorez." Anne stepped away towards the house. Her earpiece beeped.

"Red Team reporting. Target has been secured and will need medical attention. Returning to site."

She picked up her phone and made the call.

"Director, we have lost two agents. Five of our seven targets were secured, including the primary target."

"Rodriguez," he paused, his tone soft. "I am incredibly sorry about the loss of our agents. Finish up there and we can have the full debrief tomorrow. Let's get those targets to DC as soon as possible."

Anne hung up the phone. She stood there in this ornate home, wondering if what had happened was worth the cost.

BITTERSWEET

Hank stepped inside the house to wave at Niles, who was sitting on the couch with Mimi watching TV. Niles rose and walked outside to the porch.

"Niles, agents just raided the farm. The mission was a success in that all of the targets had been secured. The assessment is that you, your family, Mimi's friends, everyone is now clear of danger."

"Then why are you not smiling, Hank? What are you not telling me?"

"Niles, I don't know how to say this. There were casualties. David did not return from the mission."

Niles grabbed onto the back of the Adirondack chair beside him and lowered himself into it. His head fell into his hands. He ran them back and forth from his face and through his hair before resting them on his lap.

"He was a good person, a good agent, a good friend. This is hard news. I am not sure I am even processing it right now with everything else that has happened."

"I understand. I know you two go way back. I'm really sorry."

"Hank, please don't say anything to the kids about this. I am going to take a quick walk to collect myself. When I come back, we are going to walk in there together. I need you to deliver the news. Be elated, positive, and tell them that they are now safe, and the bad guys have been caught. I don't want to ruin their graduation day tomorrow with the reality of how this all happened."

Niles slowly and quietly walked through the dark neighborhood. The echoes of dogs barking against windows rang out nearly as constant as the crickets singing. Large picture framed windows lit with antique lamps, lace curtains, the occasional old oil lamp post at the edge of the sidewalk, now fueled by LEDs. He noticed all of these things. He smelled the humidity as much as he felt it. His senses were taking over so his mind would not.

David had been his best friend for years when they were in the service together. Working that closely with someone creates an unbreakable bond, even if they hadn't seen each other in years. He hadn't even had the chance to retire and really live his life yet. He was a truly dedicated servant to this nation. He served with honor.

Niles rounded the corner and looked at the old Queen Anne that his daughter had somehow ended up in with a group of young people that he could only have wished and hoped she would find in her life. This made him happy. David gave his life so that these kids could live in a safer world. He wouldn't have wanted to go out any other way.

Hank was still sitting on the porch when Niles walked up the concrete steps. He rose as Niles climbed the final three aged wooden steps on the landing.

"Let's go deliver the good news," Niles said as he reached for the door handle, forcing himself to smile.

START SPREADING THE NEWS

I looked up from the TV when they came in and instantly saw the smiles on their faces. I knew something had happened.

"Sweetheart, can you go gather your friends? Hank has some news to share."

Kiki was in the dining room and came into the living room. She offered to go get Sherp and Nik upstairs. I went up the front stairs to get Clara, Marcel, and Carter, who were all in their rooms during this lockdown. My mom remained on the couch with her cup of tea, watching Niles closely.

All of them came down with curious looks on their faces, wanting to know the news.

Hank began with genuine positivity in his voice, "Kids, we are happy to share that the lockdown is over and that you are safe. We have caught Mimi's kidnapper, and several other bad guys involved in their operation. All assessments say that there are no further threats, and you all are now free to do as you wish without having to look over your shoulder."

We all erupted in happiness, hugging each other, exchanging high-fives. I ran to my dad and enveloped him in a huge hug and turned to go to my mom, who had come to my side. The three of us embraced for what seemed like an eternity.

In the background, Sherp announced that this was

cause for celebration. "Nik and I are getting the beers. Hold on, y'all, this deserves a toast." And they turned and ran to the kitchen.

Everyone, including Hank, grabbed a Fat Tire out of the box.

I lifted my bottle into the air first. "To all of the agents who worked so hard to rescue me and keep us safe, thank you. And to my friends," I motioned the bottle in a circle around the room, making eye contact with each of them, "I thank you."

They all cheered and clanked bottles.

"Hold on. I am not done yet. Most of all, to my parents and my best friend, thank you for always being there for me no matter what," I paused. "Okay, now I am done. Cheers!"

It was the best-tasting beer I have had in a long time. I saw Carter across the room toasting Sherp, but he was looking right at me. I walked over to him and reached out my bottle. He reciprocated my lift and met me halfway. As the bottles clinked, I reached out for a hug. He leaned down and kissed me.

It was brief, and I think the act even caught him off guard. He slowly pulled back and looked at me. "I am really glad this nightmare is over and that you are finally safe."

I was already in a weird parallel universe and that made me think maybe I was in a dream. But then I looked over at my mom and dad who appeared to be very amused by this kiss, and I knew it was reality. Mom just raised an eyebrow and curled the edge of her mouth into a small smile. Dad just looked distracted.

I realized how exhausted I was, inside and out. I finished my beer while smiling and chatting with my roommates. The fact that graduation was tomorrow was as unreal as a Kentucky basketball fan cheering for Duke.

After a few minutes, my parents walked over to me. My mom grabbed my hand.

"Love, we are going to head to the hotel, if that is alright with you."

They looked exhausted, too. "Of course, yes. I am going to head to bed soon, too. Dad, um, do you think Hank could just double triple check that we are safe, or maybe someone can stay here tonight, just in case?"

Dad responded. "I will talk to him and see if they can keep someone here tonight if that would make you feel better. Good night, sweetheart. Get some rest. Tomorrow is a big day, if you feel like you are up to it. We will be here no matter what."

"I love you guys." In a very unlike-me fashion, I wrapped them up in another hug.

GRADUATION DAY

The whole house was awake before 9 a.m. Carter was making a big breakfast for us graduates. The air seemed lighter, and the mood in the house one of sweet relief. Nik, Clara, Marcel, and I were all graduating.

Clara walked over to the speaker and paired her phone. Bob Marley enchanted us at the kitchen with his words of wisdom. Everything was gonna be all right.

I grabbed a cup of coffee and hovered over the griddle where flapjacks were sizzling.

"Ah, a vanilla and cinnamon type of guy I see."

"Oh, not just that. Just wait. Apples, too."

"Apples in the pancakes?"

"Yes, ma'am," he said, mimicking the Kentucky twang. "My specialty."

I hadn't forgotten that Carter kissed me last night. The old Mimi would have pondered on it and thought about it for hours, playing out every scenario of why it happened and what would happen next. The new Mimi just took it for what it was. It was nice to not be riddled with anxiety and overthinking everything. What is that old saying? Something like, "experience is the best teacher?" Or maybe it is "older and wiser?" Whatever it was, my outlook on life had shifted through these last few months, and more so, these last six weeks, and even more so, the last 48 hours.

It seemed like everyone in the house was on cloud

nine today. Clara danced around the kitchen singing. Sherp stood in the corner juggling apples, providing our breakfast entertainment.

Eventually, we all parted ways and got ready for the commencement ceremony. Most of our parents were coming by the house first, around noon, for photos, and then onto the ceremony at two.

I had finished getting dressed and knocked on Clara's door as I opened it. She was scrunching her wet curly hair.

"Can you believe it? We made it."

"Shew, it was like this year was a marathon and a sprint all at the same time."

"I can't believe that 24 hours ago our lives were in danger, we were being protected by police, our phones were on lockdown, and we almost were not going to get to go to graduation."

"Everything can change in an instant. We need to be so grateful for everything we have. You know I do my daily gratitude journal. Today, I am grateful that I get to graduate college with my very best friend."

"So cheesy."

"Would you expect anything else?"

"Never. Don't ever change."

"It's almost time. Are both of your parents coming?"

"Yes, and no. My dad is coming here before; my mom will come by after. Sitting separate at the ceremony, of course. I wish they would just grow up and put the past behind them, but whatever."

"And you will be meeting Marcel's family for the first time?"

"Don't remind me. I am more nervous about that than anything else. It is nearly driving me crazy. I have scrunched my hair and changed my earrings at least ten times."

"Clara, you are the most genuine, caring, hilarious person I know. They will love you regardless of what earrings you are wearing, I promise."

She flashed me a big smile and we headed downstairs.

The rest of the day was a total blur. All of the parents at the house, the ceremony, walking across the big stage in front of thousands of people as they called my name. My two white cords indicating my honors status were flipped from one side to the other and I shook hands as I received my diploma. It was surreal. It also was loud, chaotic, busy. Absolutely one hundred percent what I needed to not remember the last few days. And it was taking its mental toll on me, the introvert.

We had canceled the big party at the house a few days ago with everything going on, which I was really happy about. It was just our parents and us hanging out now.

Clara kept feeding me drinks. I couldn't help but notice that my dad seemed a bit off, but he dropped a few bad jokes and was present, so it's not like he was totally weird. I am sure he was still trying to recover from everything that happened with me.

"I may have had one too many drinks," I said as I stumbled over to Carter.

"Whoa there, cowgirl. Need a hand?"

Sherp was observing and used that as his queue to

start clapping. He laughed and walked away.

"I am just fine, thank you. No hands needed. Just someone to protect me from Clara giving me another drink."

"Oh, a cocktail bodyguard. I can do that. In fact, I have that job on my resume if you need a reference."

"I'll have my assistant over there do the reference check later."

We both laughed. He was witty, too. Unexpected.

Nik came over. "Carter, you want to play some tunes?"

"Sure, yeah, go get your guitar. I'll go grab mine. Out back at the fire pit?"

"Sounds good. See you there." Nik turned to head through the dining room to the back stairs that led to his room.

I caught up with Clara's mom, who was a total hot mess. She meant well but it has to be hard for Clara to have to deal with her. I heard the music outside, and more than anything, I heard Kiki's singing. She had perfect pitch and an incredible tone. I invited Clara's mom to come outside and listen. She nestled in on the bench next to Clara and Marcel. They seemed so content next to each other.

Most of the parent's started to excuse themselves from the party, mine included.

"We will stop by for coffee before we head out of town tomorrow. See you then." My mom said as she gave me a hug.

"Okay, but not too early, though. Not sure how late tonight is going to go."

"We were young once, too. We understand. Take care of yourself, kiddo. Make good choices. Speaking of choices, Meems, when you want to change your hair, what choices do you have?"

I stared at him. Several drinks in, I had nothing. No comeback. Just blank.

"Do or dye."

My mom slapped him playfully across the arm. "Oh, Niles. I am sorry for your father's terrible jokes, Mimi. I will get him out of here before he tells anymore."

"Congratulations, sweetheart. We are so proud of you. See you in the morning."

I went back out to the firepit. I sat on a wooden stump and watched the guys play and Kiki sing. The orange and blue flames danced in a mesmerizing way. Everything finally seemed normal.

GRIEF AND DEBRIEF

Anne arrived in Washington late Sunday evening. She briefed the Assistant Director on Monday morning.

She sat in her hotel room looking out at the White House, knowing the walk she would take in the morning down Pennsylvania Avenue to the tall, cement building that housed the FBI offices. She stayed at the Willard whenever the federal per diem rate was available. She enjoyed the history of the hotel, and the occasional bourbon in the Round Room, which had hosted many iconic writers and thinkers over the years. Most of all she liked to know she was close to the seat of power for her country, to be reminded of what she was working to protect: America and all of her people.

However, it was on days like this, when the loss was tremendous and almost too hard to bear, that she questioned her profession. She had lost colleagues in the line of duty before, it is always a tragedy and it always gives her pause to reevaluate what she is doing. This time, it was different. David was so close to retirement. He had served loyally for many years. He was good at his job. And he was like a father figure to her. He was her mentor. Over the last decade, he had taught her a lot about judgment, instinct, patience, intelligence, and ways to approach cases. He reminded her regularly that a careful, analytical brain was her best asset, followed by her strength. This loss hurt her deeply.

She knew it must have also hurt Niles, who had worked closely with David over the years. After the raid, Anne let Niles know that the FBI sincerely appreciated his service to his country, and he could now officially retire, for the second and final time.

She knew she couldn't let the emotions she was feeling for the loss of David show through during the briefing, so she toiled through them throughout the day, composing herself and all of the findings of the case so far to prepare recommendations for next steps. This was the biggest case she had worked on, and one that would leave a mark on her career. It was a sleepless night.

Up at 0500 hours, as always. She ran the mall and monuments, returning to her routine breakfast of two scrambled cheesy eggs, whole wheat toast, and black coffee. She reviewed her files again while finishing her written report. At 8, she walked down the stairs and out to the street, turning to her right first to make a loop around the White House. She then proceeded east, down the busy pedestrian sidewalk, watching all of the suits doing the daily shuffle.

By 8:30, she had entered the imposing fortress and found her way to a remote work desk. By 9, she was in the conference room, joined by two of the headquarters' agents who had supported the mission. They confirmed the status of several outlying items. The Assistant Director entered at 10 a.m. sharp. He shook hands with all of the agents, pausing a moment longer with Anne to recognize her loss.

Anne began by handing out briefing booklets to those at the table. "Briefing on Operation Silva Stop."

"Proceed," said the Assistant Director.

"The operation has amassed dozens of missions over the last two years, beginning in February 2017 with a seizure of illegal hardwoods in Lexington, Kentucky, destined for use by private entities. That first seizure indicated that human trafficking was occurring within the same trade network.

"By August 2017, several US ports were identified as primary recipients of specific illegal timber types and operations were set up to trace the origin of that wood. By December 2017, it was confirmed that many of these seized shipments were coming from specific ports abroad with similar paperwork filings, making it seem as though it might be a syndicated effort. Additionally, during several raids, women and young girls were found, confirming that this network was trafficking people, in addition to trading in exotic woods.

"In February 2018, 15 wood shipments had been intercepted coming through the I-75/I-70 corridors in Kentucky, six of them also included humans. Several drivers became assets during this time, and, in April 2018, we confirmed that a major operator was in Kentucky and affiliated with the horse racing community. That is when we pivoted to an undercover operation to determine the identities and scope of this network."

Anne paused and looked around the room. Everyone was still with her.

"I am going to turn this over to Agent Carpenter now to talk about the international element."

Agent Carpenter rose and began speaking. "We

quickly realized that this was not an isolated set of actors. Illegal timber trade has been a major issue around the globe for decades. Given the economic and ecologic ramifications, many of the FBI's partners and analogous organizations around the world have been deeply involved and highly cooperative. As you are aware, Homeland Security and State both have significant work and interest in this area. Traditionally, this type of illegal activity was localized and for immediate benefit to whoever processed the wood. Signals in 2018 in several countries pointed to a more organized syndicate of actors. We identified Vietnam and Brazil as two key places where we were seeing unique activity and deployed agents and assets into those areas."

Anne paced back and forth by the window, watching the Assistant Director's face. Agent Carpenter turned to Anne to continue.

"In August of 2018, we recruited the help of retired Agent Niles Martin. He had worked extensively in these areas and had resources, assets, and the skillset to support this case. Plus, he was based in Louisville, Kentucky in one of the trading hubs and also had connections within the community of interest."

Carpenter continued, "With Agent Martin's help, we uncovered that the corruption in the illegal timber trade network in Vietnam, which was servicing a large part of Asia, Europe and the Middle East, was linked to an American official in the country, the Deputy Ambassador. We traced a number of overseas accounts back to him. He is in custody, and we provided the names of the Vietnamese government collaborators to our counterparts there.

"The disturbing and difficult part about Vietnam was that we found a direct correlation to human trafficking and the illegal timber trade. Villages where the trees were harvested reported missing women and children. We have determined that most of these ended up in the Middle East or Europe. A few were brought to the States and have been saved through various raids. We have identified at least 70 trafficked individuals from this region over the last two years. At least by breaking down this supply chain of timber, we are hopefully curbing that human cost for the near-term."

The Assistant Director leaned forward. "This is really excellent work. One person trafficked is one too many. Please continue."

Anne picked up a remote control off the table and turned on a screen to show an aerial photo of the Amazon. "Most of the illegally logged wood we were seeing on our route came from South America. We know this because of tree forensics. The technology here is rapidly evolving and we deployed some beta versions of new techniques that helped us pinpoint a central harvesting and trade route in the Purus River area. We knew that the company running these logs in the Amazon had been brutalizing local communities and devastating their livelihoods." She turned to Carpenter, who continued.

"It led us to a shell company and an organization that posed as a legitimate soybean, corn, and timber trading corporation. With the help of our Brazilian colleagues, we traced this corporation's origins and identified Chinese ties."

Anne interjected, "At the same time, we received credible intelligence that the primary target we had identified in Kentucky as the main operative was linked to Chinese business interests."

She nodded at Agent Carpenter to continue.

"Through this same source, we learned that prominent Chinese businessmen would be visiting Kentucky to secure new horses and conduct business. That is what set us up for this past weekend's raid. Through that raid, we believe we have completely disrupted the primary human trafficking and illegal wood network in the Midwest. Five targets were secured onsite and one more later in the day. All have been transported to a maximum-security location for further questioning."

Anne flashed the images of the six targets on the screen, walking the director through each.

"Two here are the most notable. First, Ahmad Khan, a well-respected businessman. Related to Saudi royalty. Had a small army of well-armed bodyguards and hitmen. We believe he was in charge of the South American and United States operations. Second, Li Jin Yang."

Carpenter stepped forward. "Yang was the mystery, the wild card in all of this. He was the unknown target. He owns a large agricultural conglomerate and a shipping conglomerate in China. He is rarely seen in public and keeps a very low profile. We believe he is the mastermind and the overall kingpin of the syndicate.

We are hopeful that the Deputy Ambassador to Vietnam will see it in his best interests to cooperate and, if he does, we hope that he is our link to breaking down the rest of that operation."

"We anticipate a strain on US-Chinese relations with the detainment of Yang, sir."

Rodrigues continued, "One other thing, sir. The asset who owned the nightclub is expected to make a full recovery. He played a significant role in the success of this operation. I suggest we talk with him about future opportunities. He has everything it takes to make a good agent."

The Assistant Director sat quiet for a moment. "I have already notified the State Department about the need to do some proactive diplomacy with China on this. A global illegal timber ring now broken, lives saved, and hopefully a win for the environment, too. And keep an eye on Lopez. I trust your judgment. If you think he is FBI material, then let's have a conversation with him. Excellent work, Agents, and your teams."

With that, the Assistant Director picked up the report in front of him and stood, walking towards Anne and placing a hand on her shoulder. "I am incredibly sorry about the loss of Agent Harding. David was one of the best. It was an honor to have known him."

Anne and Carpenter exchanged a few words and collected their belongings.

"See you tomorrow?" Anne asked.

"I would not miss it," he responded.

Anne left the Hoover building, dropped her belongings at the hotel, and changed back into her running clothes. She ran down Constitution to the Roosevelt Memorial bridge and through her favorite monument, Teddy Roosevelt Island. Jogging through this island forest always cleared her head. She could only think about David's memorial service that would be held tomorrow. She would give remarks along with the Director and Assistant Director. She wanted to ensure that she did her colleague the justice he deserved.

She knew the Martins were coming in, too. Niles was not fully debriefed, nor would he be. That was the last time the service would call on him, and he knew that. She felt a peace come over her in thinking about how those kids were safe now. She smiled when she remembered the spunk in Mimi, fighting off that professional wrestler thug. Anne hoped that Mimi realized how much potential she had. Anne was almost regretting her request for a transfer out of Kentucky; she had a lot of good memories there, but now it was time for something new. She made a mental note to make sure and check on Mimi when she got back.

MOVING ON

The last six weeks have been surreal. Marcel got shot, my lover got shot, my dad went all over the world tracking a case that ended up connecting back to me and my friends, I was kidnapped, fell in love twice, *and* I graduated college. I assumed I would be unable to process everything, but with each moment, it became clearer. The change inside of myself felt quantum. I felt grounded. Clara and Carter made me take up meditation and yoga with them the last few days. Before this, I would not have been caught dead doing that. Now, it felt good, like I am being true to myself in a way I have never been. I no longer feel confined by some image of myself.

Tomorrow Clara and Marcel would be moving to New York, the Big Apple.

"I need some serious extra savasana today," Clara said.

Carter smiled. "Actually, I think you need some ecstatic dance. I can feel your energy all the way over here. You need to move that body and get some of that out."

We all laughed.

Marcel's parents had taken a lot of his stuff when they were here for graduation. Clara did not have much and had sold most of her furniture over the last few days.

"So, here we are," I said, looking into Clara's eyes. "I am going to pull a Clara here and be the bad influence, taking Carter's queue. I think we should go out tonight, one last time before you move."

"In. So *in*. I thought you would never ask. I wish the Club was open. It would have been great to go back there one last time."

"Yeah, we have such a history there, but I am not so sure I would have wanted to go back. We can hit up that new place down the road, even if it is just one drink."

The whole house decided to go for one last hurrah, except Sherp because he was working. Taking two Ubers downtown, they found a few empty tables and slide them together, settling into their creaky wooden chairs.

Nik leaned in. "So, Meems, have you decided what you are going to do? Did you end up acing that LSAT of yours?"

"Oh my god, with everything going on, I haven't even checked my LSAT scores. Honestly, I don't even care anymore. I think I am going to move back to my parents for a little bit and get some stuff sorted out. Who knows? Maybe I will end up in the Big Apple, too. Or Harvard. Or somewhere totally random. Maybe Europe. Maybe Thailand."

Carter moved closer to me. "You know, you don't have to move out. If it's the rent, we can figure something out."

"Carter, it's not the rent. It's just me. I need a change of scenery and I am feeling like I need some time with my parents after everything that happened this year."

A thin, dapper-looking lumberjack type stopped by the table to grab our empty bottles.

"Y'all doing okay?" He flashed a million-dollar smile that a younger and more single Clara would have pounced all over. Unexpectedly, Kiki stepped up to the challenge. She chatted up the lumberjack with the quaffed hair in a way I have never seen her before. She followed him up to the bar to continue the conversation. Such an odd choice for her; she was so urban and edgy. I guess we all need to sometimes look outside of our box.

Clara jumped up to dance. Marcel followed her. Then Nik followed them both. Carter stayed at the table with me. He raised his eyebrows a few times, trying to entice me to go out there and "feel the vibrations," but I was not that enlightened, yet.

He leaned into my shoulder as we both looked out at the dancefloor, watching our friends bop around with huge smiles on their faces. It felt strange to be sitting here instead of in my dark corner, but I kind of liked it.

"So, I am not sure if you remember, but there was a night when things were all crazy and we were celebrating, and I kissed you. I have been thinking about that night ever since, and if I am recalling it right, it seemed like you kissed me back." Carter wasn't looking at me but he was leaning into me and looking straight ahead.

"Your recollection is correct."

"I was wondering if I could kiss you again sometime?"

"What? Here? Now?" I was slightly mortified. The

old me would have never done this. But something with Damian split me open a bit. I'd become more of a risk taker and less afraid of what people thought of me.

"Or later?"

"How about now? But not here. Why don't we head back to the house?"

Now he turned to look at me. A sly smile crept across his face. Clara turned around, beaming from ear to ear, and motioned for me to come dance. I moved my head to the right, indicating I was hanging with Carter. She gave me the Clara eyebrow acknowledgment.

The minute the door to the house opened, his arms were around me. It wasn't the type of passionate kiss that you get from lust, although you could feel that was there, too. This was the type of kiss that you waited a lifetime for and wanted to savor every moment. It was like the first sip of a good wine, the first glance off an overlook at a vista. You wanted to enjoy every second of it, not wanting it to end.

His lips were soft, his warm hands rising up through the back of my hair. My heartbeat increased. Every electron in my body radiated. This was a new and different type of connection. And I wanted more of it. I grabbed his hand and led him to the stairs. He paused.

"I am not sure I will be able to control myself if we go up there, Mimi. This is safe down here."

"Maybe I don't want you to control yourself, Mr. Jenkins. Maybe what I want is the exact opposite of control."

His body pressed against mine on the stairs. He looked down into my eyes, searching them to see if there was any hesitancy or signs of regret. I offered none.

"Mimi, I want to do right by you. I want you to respect me. As much as I want this, I think we should take this slow."

"Carter, the last two months have been us taking it slow. I do respect you. That is why I am standing here with you." I put my hand under his shirt and stroked his lower abdomen.

I felt him shudder and I grabbed his hand and continued to lead him up the stairs I took off my shirt as he shut the door behind us. He pulled his white t-shirt up over his head. We stood there for a millisecond, observing each other's bodies.

Then he closed the distance between us in two steps. His warm lips met my neck first, and then my shoulder, and then the side of my breast. I felt myself losing control. I stroked his back as he lowered his mouth to my stomach. His nose brushed my hard nipples under my bra. It sent an electric jolt through my body.

I grabbed his hair as he unbuttoned my jeans. He stopped there, with the zipper down, kissing around my abs, his hands on my hips and rested on his knees. He looked up at me. I stepped back and unsnapped my bra clasps. He rose up, slowly sliding the strap down my right arm first, freeing my right breast. He stood back and observed. I let him. I watched him as much as he watched me. It was erotic to have him undress me like this.

As he pulled the bra free, his hands came first to the small of my back, pulling me into him. My breasts pressed against his chest, and a deep kiss followed. His fingers slowly caressed the sides of my breasts. I remained pressed against him. In unison, we both took a step towards the bed.

I fumbled with the button on his shorts. He reached down and did it for me and they dropped to the floor. I wiggled out of my jeans. Our chests and lips pressed together. He lowered me onto the bed.

His hand moved up and down my thigh until I let out a whimper. I guided his hand to the edge of my lace underwear. He gently lifted the edge and made his way up and down the seam, teasing me. I reached for him and stroked him from the outside of his boxers.

Both of us were too shy to go any further. But my body couldn't take it any longer. I reached down to remove my underwear and he helped me. He slowly moved his fingers up my leg from my foot to my knee to my inner thigh and, at that moment, I felt his lips meet my breast. I grabbed his hand and then guided his fingers to the place I wanted them.

I reached for him again and used my foot to slide his boxers down. He was more of a man than I had ever seen. I caressed him and then carefully guided him to the edge of me. We played there for some time before he finally made his way inside of me. He was so gentle yet so playful. It was euphoric. Fireworks erupted inside of me. The pleasure was almost too much to bear. I never wanted this moment to end. It was the most erotic experience of my life.

We recreated that experience multiple times throughout the night, eventually falling into an exhausted sleep in each other's arms.

At some point, I slid out of his bed and tiptoed down the hallway to my room. I knew Clara would be coming in bright and early to wake me up for a final goodbye. It wasn't until I was in my own bed that I began to think about Damian.

Clara came into my room before the birds started singing. She was crying.

"Too many tears. Stop it. You are going to do amazing things. And I am going to come visit you. I promise."

"You had better. I am not sure what I will do without my Meems by my side."

"There is a thing called Facetime, you know," I said sarcastically.

"I know, but it is not going to be the same. I am going to miss you."

"You, too, and you know it." I gave her a hug then pushed out of bed, grabbing a sweatshirt, and following her downstairs.

Marcel was standing there by the door.

"Take care of her, Marcel. Push her to be as awesome as she is. Keep her bright light fueled. And make sure to give her enough of a hard time every now and then to keep her honest."

"Will do, Meems."

And that was it. They said goodbye again and the door shut behind them.

I stood there, wondering what to do next.

"Hey, you okay?"

It was Carter.

"Nope, not really, but I will be. I am headed upstairs to pack. Want to help?

"Nope, not really, but I will."

I nudged him with my elbow.

"Do you want to talk about last night?"

"Nope, not really."

Saying that felt strange.

"Let me rephrase that. Yes, I really had fun last night. I feel weird talking about it."

"Yeah, I have dreamt about that happening for a long time. I just wish you weren't moving now that it finally has."

"Carter, I am sorry if I am going to hurt you with this. I really like you. And I sincerely enjoyed last night. But right now is not a good time for me to start anything. I have a lot to figure out. I hope you can understand that with all that has happened."

He was looking at the ground, not at me. So, I kept going.

"I hope maybe one day it can work out between us. For now, at least we have something to smile about when we think about each other."

"Yeah, I understand, but it doesn't mean it doesn't hurt and that I don't wish this conversation was going a different direction right now. I am going to go get ready for the day. Are you good here or do you need my help?"

"I am good. Thanks. And I am sorry."

He walked out of the room. Tomorrow, my parents would be coming to collect my stuff and move me back to Louisville. Today was my last day in this house and in Lexington.

I decided to walk to The Bean for one last cup of coffee. I was headed down the stairs when Carter unexpectedly came out.

"Mind if I join you?"

"I would like that very much."

"I didn't want you moving out with things being all weird, so figured we could walk as friends to get a cup of coffee."

We both smiled and walked down the chipped and broken sidewalk, past the old houses on the edge of campus, listening to the bird's sing. Carter told me about his thesis progress.

I was really going to miss The Bean. The routine, the people watching, my apple table. Hopefully, wherever I end up, I can find a coffee spot as good as this one.

Back at the house, I worked to get packed. My phone rang and I saw Anne's name pop up. My heart dropped. Why would she be calling me? I answered.

"Hello?"

"Hi Mimi, it's Anne. I was wondering if you might have a chance to connect before you leave town?"

"Actually, Anne, I am moving tomorrow. I'm almost done packing, if you wanted to meet now."

"Great. Let me treat you to lunch. Where do you like to eat?"

"Oh, there is this great Mediterranean place over on the east side of campus."

"I know the one. See you there at noon."

"Sounds good."

Time flew and before I knew it was 11:30. I pulled on my shoes and headed out the door to meet Anne, who was already there and waiting for me in one of the booths.

We caught up and made small talk at first.

"So, Mimi, I know you went through a lot this past few months. When agents go through things like that, some of them struggle. You have probably heard of PTSD. It is a bit like that, or it *is* that. I just wanted to make sure you were getting help if you needed it. I have resources for you."

"Actually, I am surprisingly good. That whole complicated mess helped me evolve a bit. Not saying I wouldn't benefit from a therapy session or two in the future, but so far, so good."

"I also wanted to talk to you about your future. What are you thinking about doing next?"

"Oh, well, probably law school, but I am having a bit of a 'who do I want to be when I grow up?' crisis at the moment. So, not exactly sure."

"I understand that. The good news is, you have time, and you can do lots of things in your life."

"People keep telling me that. It doesn't feel that way, though. It seems like once you make a decision, you are locked into a career to help pay off the investment you made getting there."

"Yes, I guess that is the reality for most people. Listen, Mimi, I want to encourage you to keep up with your training. I watched the videos of you fighting off that big guy at the Club and then trying to escape the guy at the concert. I am also aware that you keenly put together quite a few puzzle pieces of what we were working on, without any assistance. Those are unique skillsets that not many people have."

"What are you trying to say, Anne?"

"I am trying to say that an organization like mine could really benefit from getting someone like you as part of our team."

"Whoa, the FBI? Yeah, but no thanks. I mean, I am so not qualified for that stuff."

"I would say you are more qualified than you think. And I also think you would make an incredible lawyer. I think you will be successful in whatever you choose to do."

"Anne, um . . . I have been wondering about Damian, do you have any information on how he is doing?"

"Would you like to go see him for yourself?" she asked.

I sat there for a few moments, not sure if I wanted to see him again. After everything that happened, I was ready to move on, to close this chapter of my life book.

"Actually, I am not sure I am ready for that, but could you pass along a letter to him?"

"Of course."

While she drank her coffee, I pulled out my notebook and began to write. It came easy.

Damian, for everything that has happened to you, I am sorry. Ends up I am even stronger than I thought I was and it appears you are, too. Anne let me know that you will make a full recovery, this is really great news. I am not sure if you will go back to Miami or somewhere else, but I am guessing you are going to get as far away from this place as possible and I don't blame you. There are not really words to describe our short time together, but I am grateful for it. I hope one day you can find peace around your mother and brother. Take care and be safe. Your friend, Mimi

I ripped the paper out of my notebook and folded it. Anne glanced up from her coffee and I handed it across the table to her. She placed it in her bag.

We finished up our meals and said our goodbyes. I walked home pondering what Anne had said, what Damian would end up doing, and what my next steps were.

SOME DOORS CLOSE, OTHERS OPEN

My parents showed up around 9 the next morning. My remaining roommates were there to help move my furniture out to the truck. They had plenty of space in their garage for me to store it. It seemed surreal that this phase of my life was over. My mom ordered us all pizza and we ate and said our goodbyes.

Saying farewell to Kiki was especially hard. She was special. She helped save my dad. She was one of the most interesting people I had ever met. I knew we would be friends forever. And she agreed. We would be in frequent contact.

Sherp and Nik gave me hugs and high-fives and wished me the best. I hoped they would last. I don't know, though. Nik has big plans and Sherp just seemed stuck in his ways of being a dirtbag climber. I guess only time would tell.

Carter followed me up the stairs to grab my last few items.

"Promise me you will follow your heart."

"What is that supposed to mean?"

"It means that I don't want you to get trapped into what you think you should do or who you should be. You are amazing just the way you are, okay? I know you are going to do incredible things. I only hope that you will send me a text or give me a call every once in a while."

"Carter, you have come to be one of my closest friends during one of the hardest times in my life. I won't ever forget how you have been there for me. I will for sure keep in touch. And I hope you do the same."

We hugged. He leaned down and gave me the most loving, emotionally connected kiss I have ever had.

I walked back downstairs and saw Mom and Dad waiting at the door.

I said goodbye to the beautiful house and these people who had become my chosen family one last time.

We got into the car, and I looked out ahead to what my future could be.

CHEMICAL SIGNALS

Each person is unique. While our DNA is nearly exactly the same as those around us, we are each our own person. Humans put off chemical signals through their pheromones when they are attracted to another. These signals often determine the success of a relationship. Through chemical signals and DNA transfer, we remain a viable species.

EPILOGUE

The paper sat on his lap as he looked out the window of the airplane. He had carried it unopened waiting for the right moment to read it. He already knew it was a final goodbye. There was no turning back. Too much had happened. They both needed to move on from those few weeks of chaos. Glancing down, he saw the lights of the National Mall running all the way to the Washington Monument. The wing of the plane dipped, indicating they were coming in for landing. He unfolded the piece of paper. It was the first time he had seen her handwriting. It was as meticulous as she was. The attendant came around asking one last time for trash and he crumpled the letter as he placed it into the bag. Damian was ready to leave his past behind and move into his new future.

ACKNOWLEDGMENTS

This book is dedicated to my family. I am forever grateful for the unconditional love and support of my parents and brother. My husband has encouraged me every step of the way and never once complained about the countless hours I spend with my characters. My world is made brighter every day by my son. His inspiritment kept me going and I hope I make him proud. Big hugs go out to my beta readers. Their honest feedback helped deepen and improve the book. And for all of my friends and family that nudged me to put pen to paper, that held me up, that motivated me along the way – you know who you are – thank you!

A few thoughts about the critical topics covered in this book – deforestation, illegal logging, and human trafficking. Deforestation is one of the great environmental challenges of our time and sustainable forest management one way to address it. Trees are incredible! Our world's forests provide us with so many goods and services. They help keep our air clean and water fresh, provide respite for our soul and body, and give us so many other benefits. Trees can be a renewable resource and products made with wood continue to store carbon long into the future. Our forests are part of the solution for addressing climate change. I encourage readers to learn more about the benefits of sustainable forest management. Consumer choices for certified sustainable products and voicing concern to policy makers for tackling illegal timber trade can help shift supply chains. Addressing these issues is critical for the future of our planet. Also, for those that do not know, as I did not before researching this book; the United Nations says human trafficking is the "second largest and fastest growing criminal industry in the world." More awareness is needed globally to understand how to identify this exploitation, where it is happening, and what we can do to help organizations supporting people that are able to escape it.

I passionately believe science can make our world a better place. To the researchers, scientists, environmentalists, law enforcement officers, forest managers, and all those working in service to prevent deforestation and enhance sustainable forest management; and to those working to stop human trafficking and help victims of this terrible fate – please keep doing what you are doing and know you are appreciated!

ABOUT THE AUTHOR

J.C. Rivers is a Kentucky-rooted, Colorado-based author.

Rivers uses science and suspenseful storytelling as a lens for today's environmental issues. When she is not writing, she can be found in the mountains or on the water.

Learn more about her work at http://jcriversofficial.com/ and on Instagram @j.c._rivers.

Made in the USA
Middletown, DE
23 August 2023

37249926R00217